The Rise and Impact of Rare-Earth Permanent Magnets

Georgia

Contents

1 Introduction

The early attempts in rare-earths-based permanent magnets development date back to 1959 when Nesbitt *et. al* [1] (followed by Hubbard *et. al* in 1960 [2]) reported high coercivity for $GdCo_5$ associating it to the large magnetocrystalline anisotropy of the compound. However, only in 1966 when Hoffer and Strnat [3] reported on YCo_5 did the scientific community recognize that the $RECo_5$ (RE - rare-earth element) compounds offered a fertile new ground for further investigation. This lead to the discovery of $SmCo_5$ as the ideal combination and a new era began in the research for novel permanent magnets. In the magnetic applications of cobalt, Alnico was the alloy of choice before 1978. The Sm-Co magnets were already being developed and showed great promise in replacing Alnico due to their superior magnetic energy densities. However, the cobalt market experienced a major disruption in the late 70s (the "Cobalt Crisis") triggered by the political instability in Zaire (now Democratic Republic of Congo), the main supplier to the Western world. Prices spiked and actors on the entire supply-chain were severely affected, from upstream raw materials producers to downstream component and product manufacturers. $Nd_2Fe_{14}B$-based magnets were formally introduced in 1983 at the Conference on Magnetism and Magnetic Materials (MMM) in Pittsburgh, USA [4]. They were simultaneously invented by Sagawa *et al.* [5] at Sumimoto Special Metals who used the powder metallurgy and sintering approach and by Croat *et al.* [6, 7] who developed the rapid quenching processing route at General Motors. Having the more abundant Nd and Fe as main components, Nd-Fe-B permanent magnet alloys were successfully developed with the aim to reduce raw material cost and ultimately replace the more expensive Sm-Co magnets. Material substitution has been proved to be an efficient strategy taken in response to availability-driven price increases as it can reduce the demand for scarce resources. The industry responded fast by substituting Sm-Co alloys particularly in applications with limitations on weight, size and energy. $Nd_2Fe_{14}B$ alloys achieved the best performance compared to any previously designed hard magnetic material in terms of intrinsic coercivity (H_{ci}), remanent magnetization (B_r) and energy density product $|BH|_{max}$, the most common figure of merit used for permanent magnets evaluation. Magnetic properties improved continuously so that, currently, the energy product of commercially available sintered $Nd_2Fe_{14}B$ magnets exceeds 400 kJ/m^3 while the highest value (474 kJ/m^3) was attained by Hitachi as reported by D. Harimoto in 2007 [8]. Over the years, the applications integrating $Nd_2Fe_{14}B$ magnets increasingly diversified, ranging from various types of electric motors (servomotors, synchronous motors, spindle and stepper motors, electrical power steering, etc.) to hard disk drive voice coil motors (VCMs), acoustic equipment (microphones, loudspeakers for electronic gadgets like smartphones and laptops, headphones) and magnetic resonance imaging (MRI) instruments. Since the launch of Toyota Prius in 1997, the use of $Nd_2Fe_{14}B$-based magnets has become common for emerging green transportation technologies. They are vital components in traction motors for hybrid electric vehicles (HEV, such as Prius) and fully electric/battery electric vehicles (BEV, like Nissan Leaf, launched in 2010). Furthermore, renewable energy technologies like direct drive wind power generators rely heavily on the use of $Nd_2Fe_{14}B$-based permanent magnet alloys. The performance of $Nd_2Fe_{14}B$-based alloys is known to depend strongly on temperature.

Their coercivity has negative temperature coefficients, dropping significantly with increasing temperature which restricts the operation conditions. Motor applications require high stability of coercivity in demagnetizing fields at elevated temperatures. This can be achieved by alloy composition modification with additives like Dy which increases magnetic anisotropy, ensuring superior coercivity values at high temperatures [9].

In 2011, China, the world's largest producer of REEs (more than 95 % of REEs raw materials production), drastically reduced the REEs export quota as part of a strategy to expand the internal permanent magnet production. As a consequence of low supply and market monopoly (China is the only industrial scale exporter of RE metals), the prices of Nd and Dy spiked dramatically, going up even more than tenfold as was the case of Dy, generating a full-fledged crisis on the REEs market [10]. In this context, Nd, indispensable for $Nd_2Fe_{14}B$-based permanent magnets, and Dy, necessary alloying additive for high temperature applications, substantially increased the magnet production costs. They were listed as highly resource-critical, strategic raw materials (Figures 1.1 and 1.2), their high fluctuations-prone supply severely affecting both the european and american energy and automotive industries [11, 12].

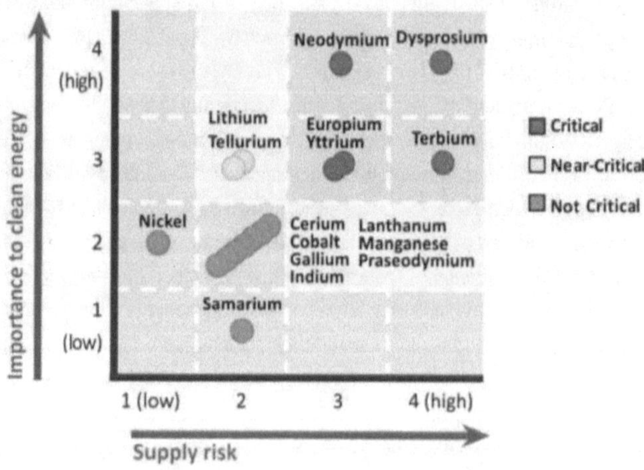

Figure 1.1: Medium term (2015 - 2025) raw materials supply criticality matrix [11].

Moreover, correlated with an increasing demand for associated consumer goods, the projected demand for REEs is estimated to grow [13]. Therefore, in order to insure stability, besides performance, important factors like cost, resource-efficiency and sustainability need to be addressed and the production/development of $Nd_2Fe_{14}B$ magnets should be adapted considering the high potential fluctuations on the rare-earths market.

In the aftermath of the rare-earth supply disruption, the availability of rare-earth elements from primary resources has come into question and the industry, policy makers and the research community made efforts for a coordinated response. Research to build know-how in material substitution (but also on process and product substitution) is essential for progress and maintaining flexibility and adaptability to changing markets. Different strategies have been evaluated as potential solutions to alleviate the crisis: (i) recycling of end-of-life devices extracted magnets by technologies that are typically used in production [14, 15, 16]; (ii) reduction of heavy rare-earth content (i. e. Dy) in $Nd_2Fe_{14}B$ magnets by grain boundary diffusion process (GBDP), a technology which implies the targeted use of the expensive performance-enhancing rare-earth additive [17]; (iii) development of Dy-free $Nd_2Fe_{14}B$ magnets and substitution of Nd by cheaper alternatives [18]; (iv) development of rare-earth-free magnets [19].

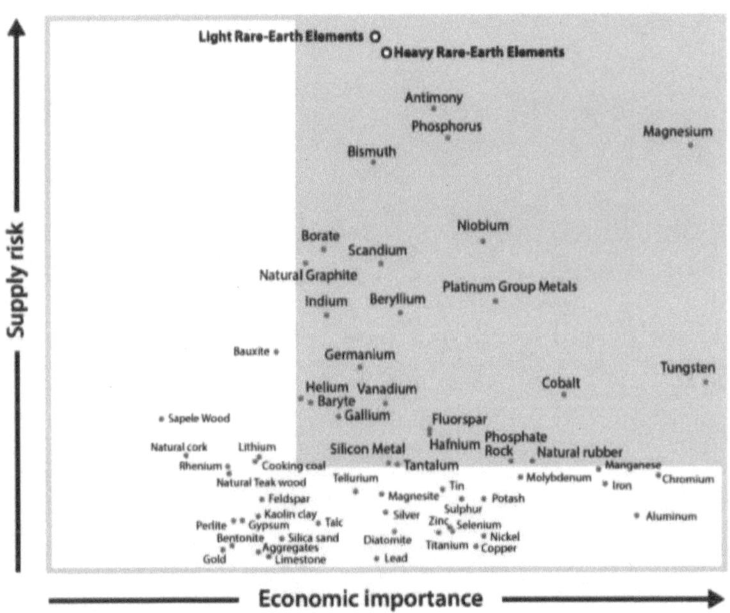

Figure 1.2: Chart of critical (blue area) and non-critical (outside the blue area) raw materials according to the 2017 report of the European Commission placing the heavy and light rare-earth elements at the highest supply risk [20].

Nd occurs in monazite and bastnäsite ores together with Ce and La but in smaller proportion in comparison and large amounts of Ce and La are stockpiled after separation as the demand for Nd is significantly higher [20]. The supply/demand imbalance within the monopolistic rare-earths market context made Nd highly critical and drew attention on Ce and La as much cheaper potential substitutes in permanent magnet production.

The sharp increase of rare-earths prices in 2011 refocused research efforts particularly towards the development of Ce-based permanent magnets which were then introduced in production in China in 2015 [21].

This work deals with rare-earth substitution in $Nd_2Fe_{14}B$-based alloys aiming for cost reduction and balanced use of raw material resources in $Nd_2Fe_{14}B$-based permanent magnets fabrication. Cerium and lanthanum were the elements of choice used as substitutes for Nd and, starting from rapid solidification as synthesis approach, two alloy processing routes were adopted in magnet preparation: (i) melt-spinning followed by hot-pressing and hot-deformation; (ii) strip-casting followed by hydrogen decrepitation and hydrogenation-disproportionation-desorption-recombination (HDDR).

2 Background

2.1 Fundamental aspects of magnetism

James Clerk Maxwell unified and derived concepts of electricity, magnetism and optics into what is known as Maxwell's equations, the foundation of classical theory of electromagnetism. The four equations (Gauss's laws for electric and magnetic fields, Maxwell-Faraday equation and Ampère's law) are the mathematical distillation of decades of experimental observations that linked magnetic phenomena to electricity [22]. The fundamental ideas behind them imply that a magnetic field is produced by electric charges in motion and time-varying electric and magnetic fields generate each other and can propagate indefinitely through space. Magnetism is therefore related to the electronic structure of matter and arises from the orbital motion of the electron around the nucleus and also by the electronic spin motion. A magnetic force/field (termed as magnetic moment) is associated with the electron which behaves as a magnetic dipole. The magnetic dipole moment of an electron is caused by either its orbital or spin angular momentum and is expressed by a physical constant called the Bohr magneton (μ_B):

$$\mu_B = \frac{e\hbar}{2m_e} = 9.27 \times 10^{-24} \, A \cdot m^2 \, (SI) \tag{2.1}$$

where e is the elementary charge, $\hbar$ is the reduced Plank constant and m_e is the electrons rest mass. Electrons which are paired in orbitals have their magnetic spin moments cancelled out (m_s, the electron spin magnetic quantum number is either $+1/2$ or $-1/2$). The valence shells and the partially filled, inner 3d orbitals of transition elements and 4f orbitals of rare-earth elements contain un-paired electrons which give rise to a strong magnetic moment. For Fe, Co and Ni the magnetic moment is supported only by the electronic spin motion while in the case of some rare-earths atoms both orbital and spin motion contribute to the magnetic moment.

2.1.1 Types of magnetism

Electric fields can induce electric dipoles in materials and, if the atoms or the molecules themselves are permanent electric dipoles, an external electric field will tend to align them and the degree of success depends on electric field strength and on temperature. At low temperatures (less thermal agitation) the dipoles will be more easily aligned. Similarly to electric fields, an external magnetic field can induce magnetic dipoles in materials, at the atomic scale.

The degree of magnetization of the material (M) depends on the applied magnetic field (H) and on the magnetic susceptibility (χ_m):

$$M = \chi_m H \qquad (2.2)$$

Susceptibility (χ_m) is a dimensionless proportionality constant which is material specific and determines the response of the material to an applied magnetic field. Depending on their magnetic susceptibilities, materials can be classified as *diamagnetic*, *paramagnetic* or *ferromagnetic*, *antiferromagnetic* and *ferrimagnetic*. Concepts of antiferromagnetism and ferrimagnetism are not detailed in this work. The above classification can also be related to the distribution of magnetic moments which determines magnetic properties and the behavior of a material in a magnetic field (Figure 2.1).

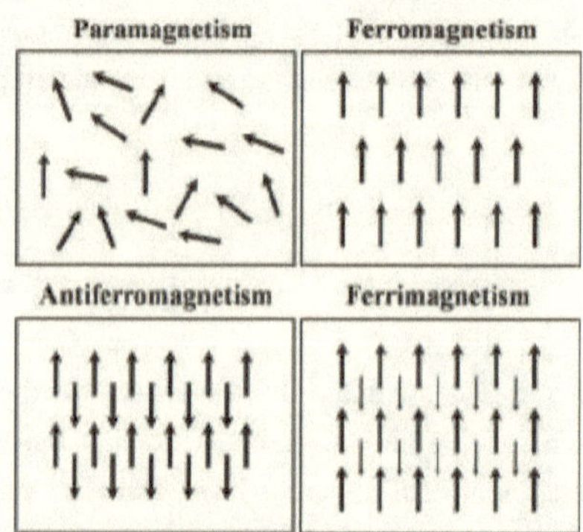

Figure 2.1: Relative orientation of electrons spins: randomly aligned in paramagnetic state, aligned parallel in ferromagnetic state, anti-parallel with equal magnitude in antiferromagnetic state and anti-parallel with different magnitudes in ferrimagnetic state (summarized from ref. [22], details in Chapter 4.2).

Diamagnetism

Upon exposure to an applied external magnetic field all materials will to some extent oppose it, and this effect is called diamagnetism. An anti-parallel electromagnetic force is generated on atomic scale, opposing the external field. The magnetic field inside the diamagnetic material is smaller than the external field because the dipoles oppose the external field. This effect is extremely weak and it can be masked by other magnetic phenomena in the material. Diamagnetic substances have filled orbital shells resulting in a zero net magnetic moment, so in the absence of an external field this effect will not occur. Their magnetic susceptibilities (χ_m) are negative (with magnitudes of -10^{-5} to -10^{-6}), independent of temperature and very small in magnitude compared to paramagnetic materials and negligible compared to ferromagnetic materials.

Paramagnetism

In many materials the atoms or the molecules carry a permanent magnetic dipole moment. In the absence of an external magnetic field, the dipoles in the paramagnetic materials are randomly oriented, cancelling each other so the net magnetic field inside the material is zero. When an external magnetic field is applied the dipoles will tend to align in the direction of the field (depending on the strength of the field and the temperature). Magnetic susceptibility (χ_m) in this case is positive (in the order of 10^{-3} to 10^{-5}) and depends on temperature according to the Curie law:

$$\chi_m = \frac{C}{T} \tag{2.3}$$

where C is the Curie constant per gram and T is the absolute temperature.

Paramagnetic materials are attracted toward magnetic fields while diamagnetic materials are repelled from the field because in diamagnetic materials the internal induced field opposes the external field whereas the paramagnetic state supports it.

Ordered magnetic states - ferromagnetism and magnetic domains

A particular case of paramagnetism is ferromagnetism [23, 24, 25, 26]. The atoms in the ferromagnetic material have permanent magnetic moments, but in this case they are "grouped" in domains (Weiss domains) which represent magnetically saturated areas whereby the moments are all spontaneously (that is in the absence of an external field) aligned in the same direction. Adjacent domains are separated by boundary regions (Bloch walls) across which the direction of magnetization gradually changes. The domains are uniformly distributed throughout the ferromagnetic material, cancelling each other so there is no net resulting magnetic field. The formation of magnetic domains is a mechanism of magnetic energy minimization. Splitting into magnetic domains minimizes the magnetostatic energy associated to the demagnetizing field that would be generated around a single-domain ferromagnet as shown in Figure 2.2.

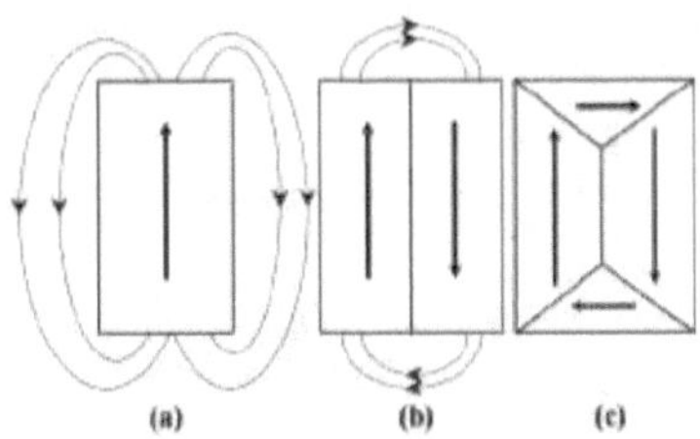

Figure 2.2: Schematic showing how splitting into domains can reduce the demagnetizing field therefore reduce magnetostatic energy (a - uniformly magnetized crystal; b - split into two domains; c - closure domain structure) - adapted from ref. [23].

The parallel orientation of the magnetic moments of neighboring atoms in ferromagnetic crystals is supported by interatomic interactions between electrons (or exchange coupling). The exchange interaction energy for two atoms i and j with spin angular momentum $S_i\hbar$ and $S_j\hbar$, was put in a mathematical model by Heisenberg in 1928 and is written as:

$$E_{ex} = -2J_{ex}S_iS_j = -2JS_iS_j\cos\varphi \tag{2.4}$$

where J_{ex} is the exchange integral and φ is the angle between the two spins. For ferromagnetic coupling the exchange integral J_{ex} is positive, the exchange energy E_{ex} having a minimum if the spins are parallel ($\cos\varphi = 1$). The coupling between neighboring magnetic moments in paramagnetic materials is very weak compared with the thermal energy. In ferromagnetic materials the exchange coupling between neighboring magnetic moments is very strong. Susceptibility (χ_m) is positive (can be very high in magnitude, in the order of 10^1 - 10^4) for ferromagnetic materials and is described by the Curie-Weiss Law:

$$\chi_m = \frac{C}{T - T_C} \tag{2.5}$$

where C is the Curie constant per gram and T is the absolute temperature and T_C is the Curie temperature.

All ferromagnetic materials become paramagnetic above a critical Curie temperature (T_C). The Curie temperature is an intrinsic property of the material, being the point where the thermal agitation is high enough to break the cooperative spin ordering. The exchange energy (therefore also the Curie point) is sensitive to the interatomic distances and the diameters of unfilled shells. The Bethe-Slater curve is a plot of the interatomic spacing/3d shell diameter ratio versus the exchange integral (Figure 2.3). It indicates that Co has the highest Curie temperature among the ferromagnetic elements.

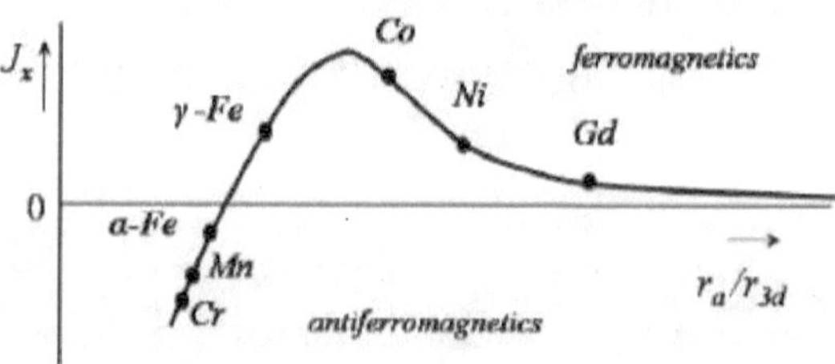

Figure 2.3: Bethe-Slater plot of the exchange integral (J_{ex}) versus the interatomic spacing/3d shell diameter ratio (r_a/r_{3d}), adapted from ref. [24].

2.1.2 Magnetocrystalline anisotropy

Magnetocrystalline anisotropy is an intrinsic property and it arises from the difference in energy of the magnetization with respect to the axes of a particular crystal. It is a result of spin-orbit-lattice couplings and it locks the spontaneous magnetization on a particular, energy minimizing crystallographic direction. Within a ferromagnetic domain the favorable crystallographic direction in which the energy is minimized is called *easy direction* (or the *easy axis* of the crystal). The energetically non-favorable direction is the *hard axis*. For non-cubic crystals (tetragonal and hexagonal symmetry, which are most relevant for permanent magnets) with uniaxial anisotropy the magnetocrystalline anisotropy energy density is written as:

$$E = K_1 \sin^2 \theta + K_2 \sin^4 \theta + ... \tag{2.6}$$

where θ is the angle between the magnetization vector and the anisotropy axis (the c-axis). K_2 relates to the planar component and is negligible for the uniaxial crystal.

The magnitude of K_1 (measured in J/m^3 (SI)) anisotropy constant is temperature dependant and it indicates the energy density required to switch the orientation of the magnetization from the easy axis to the hard axis of the crystal. Alpha-Fe (BCC) possesses $K_1 = 1.5*10^4$ J/m^3 and $K_2 = 5*10^4$ J/m^3, for Co (HCP) $K_1 = 5*10^5$ J/m^3 while for the hard ferromagnets SmCo$_5$ (hexagonal) and Nd$_2$Fe$_{14}$B (tetragonal) K_1 reaches $2*10^7$ J/m^3 and $5*10^6$ J/m^3 respectively [27].

The Figure 2.4 contains a chart of the K_1 constant versus the saturation magnetization $\mu_0 M_s$ for representative magnetic materials with uniaxial magnetocrystalline anisotropy (or shape anisotropy in the case of Alnico type alloys) and Curie points high above room temperature. Regions for hard, semihard and soft magnets are delimited on the plot relative to the lines corresponding to the magnetic hardness parameter [28] which is defined as:

$$\kappa = \sqrt{(K_1 / \mu_0 M_s^2)} \tag{2.7}$$

where M_s is the saturation magnetization and μ_0 is the vacuum permeability.

The anisotropy constant K_1 can be calculated from the anisotropy field (H_a) by the following formula:

$$H_a = \frac{2K_1}{\mu_0 M_s} \tag{2.8}$$

The anisotropy field (H_a) is defined as the intersection of the magnetization curves measured parallel and perpendicular to the easy direction of the crystal (shown for Nd$_2$Fe$_{14}$B single crystal in Figure 2.5). The saturation magnetization (M_s), or spontaneous magnetization for single domains, is an intrinsic property of the material showing the maximum degree of magnetization reached when the spins are all aligned in the direction of the applied field. For the saturation magnetization (M_s) on the easy axis to be reached, a low applied magnetic field is sufficient whereas for the hard axis to saturate a high magnetic field is necessary.

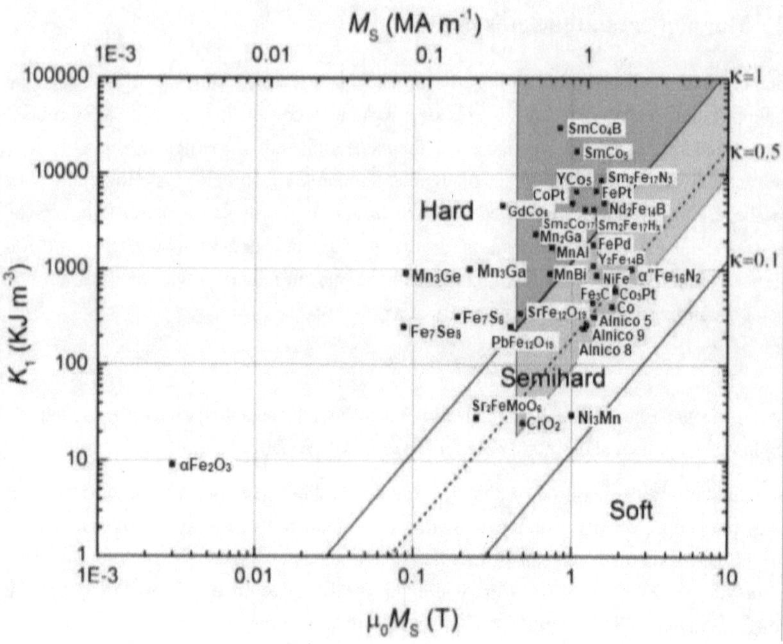

Figure 2.4: Anisotropy constant K_1 vs. saturation magnetization $\mu_0 M_s$ plot for materials with uniaxial anisotropy. Hard, semi-hard and soft magnetic materials are highlighted and the solid and dotted lines correspond to the magnetic hardness parameter $\kappa = 0.1$, 0.5 and 1 [28].

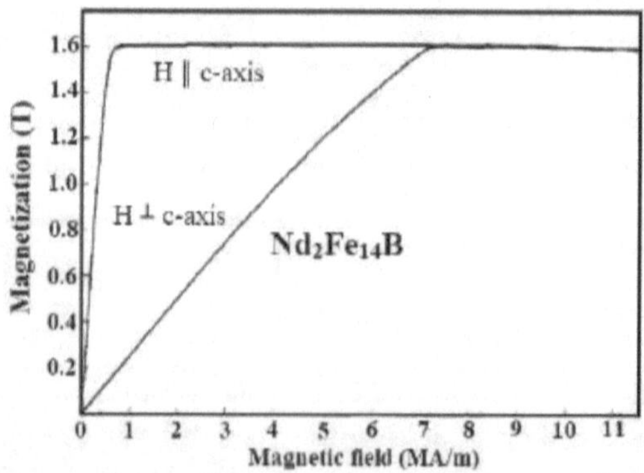

Figure 2.5: Room temperature magnetization vs. magnetic field curves of a Nd$_2$Fe$_{14}$B single crystal [29].

2.1.3 Hard ferromagnets and magnetic hysteresis

An ideal magnet exhibits rectangular hysteresis with the saturation magnetization (M_s) remaining unchanged both in zero field and in reverse field until the coercivity point is reached. It is based on the model Stoner-Wohlfarth, calculated for an ellipsoid sample, assuming that magnetization reversal is uniform, realized by coherent rotation of the magnetic spin moments in the direction of the field [23]. The hysteresis behavior of real ferromagnets originates in the motion and growth of the magnetic domains and is a function of domain wall mobility which is controlled by the microstructure. When an external field is applied the domains align in the direction of the field. Upon magnetization, the domains with alignment in the direction of the applied field grow at the expense of the misaligned ones and, in high fields, the domains rotate to align with the field.

The processing history of the sample - the size, shape and orientation of the hard magnetic phase grains, as well as the nature and distribution of any secondary phases that are present will impact the hysteresis loop. Microstructure dependent, extrinsic magnetic properties (besides these, in principle also the saturation magnetization M_s) are derived from the hysteresis curve.

Figure 2.6 contains schematics of M(H) and B(H) hysteresis loops of an ideal magnet (a) and a typical M(H) loop for a real magnet (b). The M(H) and B(H) loops relate by the formula:

$$B = \mu_0 (H + M)$$
(2.9)

where B is the magnetic induction or magnetic flux density (measured in T (SI)).

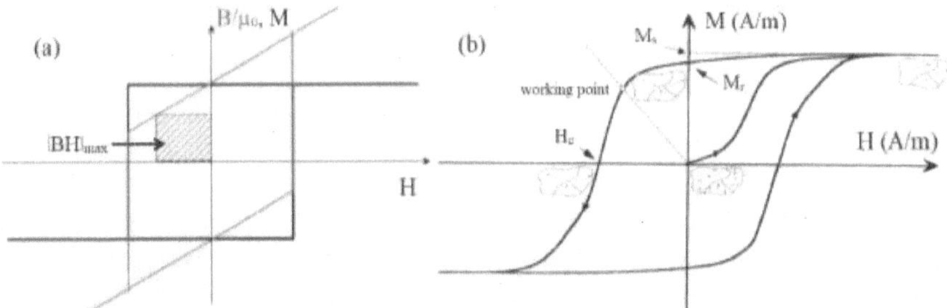

Figure 2.6: (a) M(H) and B(H) hysteresis loops of an ideal magnet; (b) M(H) loop of a real magnet. H_c is the intrinsic coercivity (internal reverse field that reduces M to zero); M_s is the saturation magnetization. M_r is the remanent magnetization at H = 0. The dotted line intersecting the loop at the working point is the load line, with the slope 1/N, where N is the demagnetization factor which depends on the magnet geometry [30].

Coercivity is the capacity of the ferromagnet to resist an external magnetic field operating in the direction opposite to the orientation of the magnetization. It is expressed as intrinsic coercive field (H_c), the value of the externally applied magnetic field that brings magnetization (M_r) to zero. The intrinsic coercivity on the ideal loop is equal to the anisotropy field.

For a real magnet however, coercivity is only about 20 to 40 % of the anisotropy field. Coercivity development is governed by three distinct magnetization reversal mechanisms involving *coherent spin rotation* for a single domain structure (the ideal model), *domain wall nucleation and growth of reverse domains* and *domain wall pinning* for multidomain structures (real magnet). Coercivity increases with decreasing particle size reaching a maximum when the particles attain the single-domain critical size (illustrated for $Nd_2Fe_{14}B$ magnets in Figure 2.7). Here, their behavior follows the Stoner-Wohlfarth coherent spin rotation model. Below the critical single-domain size (toward a few nanometers) the particles tend to the superparamagnetic state at room temperature so coercivity vanishes. Above the critical domain size, the competition between the increasing magnetostatic energy and the domain-wall energy favors domain-wall formation and the single-domain particle splits into multi-domains which leads to a decreasing coercivities.

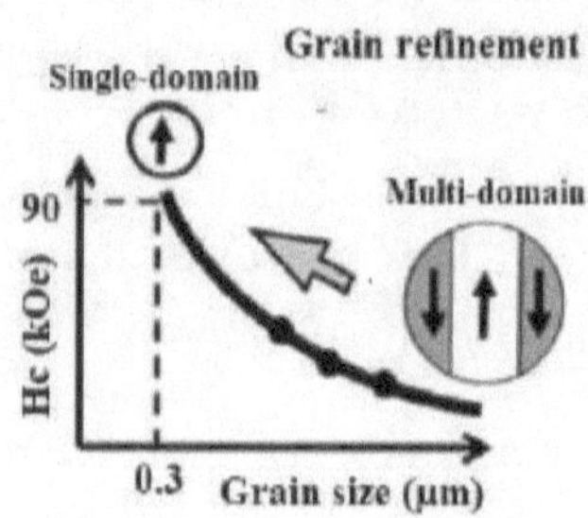

Figure 2.7: The evolution of coercivity as a function of grain size, redrawn from ref. [31].

The coercivity of *hard magnets* exceeds 400 kA/m while *soft magnets* have coercivities below 10 kA/m, intermediate values being associated with *semihard magnets* (mainly used as magnetic recording media).

Remanent magnetization M_r (or *remanent induction/ remanent flux density* $B_r = \mu_0 M_r$ on the B(H) curve) is the value of magnetization remaining in zero field after the sample is saturated. Its magnitude is dependent on the magnetic anisotropy and exchange interaction strength and on the processing condition of the material (induced shape anisotropy, texturing).

The stored magnetic energy by unit volume expressed by the *maximum energy density product* $|BH|_{max}$ (kJ/m^3), is the one of main criteria in the assessment of permanent magnets performance. It is given by the maximum rectangular area beneath the B(H) demagnetization curve (Figure 2.6 shows the maximum theoretical $|BH|_{max}$).

2.2 The Nd$_2$Fe$_{14}$B compound and related R$_2$Fe$_{14}$B isomorphous structures

The Nd$_2$Fe$_{14}$B crystal lattice has tetragonal symmetry (space group P4/*mnm*) and the unit cell contains 68 atoms (4 formula units) configured in a open-frame repeating layer structure as shown in Figure 2.8. Within the unit cell, six distinct Fe sites, two rare-earth and one B positions can be distinguished [32].

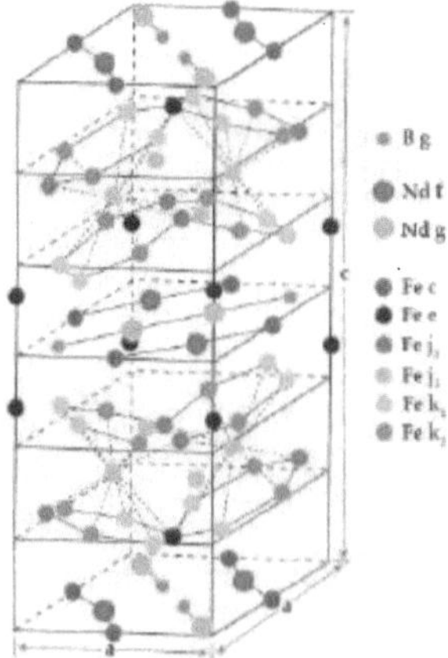

Figure 2.8: Nd$_2$Fe$_{14}$B tetragonal unit cell structure (having different coordination environment, Nd and Fe atoms are differentiated according to their site occupancy) - *c* axis is not to scale [32].

The outstanding magnetic properties of the Nd$_2$Fe$_{14}$B compound arise from the exchange coupling between the large spin magnetic moments of the transition metals (Fe, Co) and the magnetic moments of the rare-earth component. The high magnetocrystalline anisotropy is provided by the interaction of the spin magnetic moments with the Nd 4f electron orbital magnetic moments, which stabilizes the magnetization direction with respect to the crystal axes by electric field effects [32].

It was found that all rare-earth elements (except for Eu and Pm) and also Y and Th form crystal structures isomorphous with the Nd$_2$Fe$_{14}$B tetragonal structure [33]. The lattice *c* parameter of the R$_2$Fe$_{14}$B compounds shows a decreasing trend with increasing atomic number Z in the lanthanide series. This behavior is related to the lanthanide contraction effect that is the decreasing atomic radii of the trivalent lanthanide ions the from La ($Z = 57$) to Lu ($Z = 71$). The intrinsic magnetic properties (saturation magnetization - M$_s$, anisotropy field - H$_a$, and Curie transition temperature - T$_C$) of the R$_2$Fe$_{14}$B compounds fluctuate dramatically within the lanthanide R component series. The room-temperature lattice constants and intrinsic magnetic properties of the R$_2$Fe$_{14}$B compounds are listed in Table 2.1.

Table 2.1: Room temperature lattice constants a and c for selected tetragonal $R_2Fe_{14}B$ compounds and intrinsic magnetic properties (saturation magnetization - M_s, anisotropy field - H_a, and Curie transition temperature - T_c) - data centralized from [33].

Compound	Lattice constants		Intrinsic magnetic properties			
	a (Å)	c (Å)	M_s (μ_B/f. u.)	$\mu_0 M_s$ (T)	$\mu_0 H_a$ (T)	T_c (°K)
$La_2Fe_{14}B$	8.82	12.34	28.4	1.38	2.00	530
$Ce_2Fe_{14}B$	8.76	12.11	23.9	1.17	2.60	424
$Pr_2Fe_{14}B$	8.80	12.23	31.9	1.56	7.50	565
$Nd_2Fe_{14}B$	8.80	12.20	32.5	1.60	7.30	585
$Tb_2Fe_{14}B$	8.77	12.05	14.0	0.70	~ 22	620
$Dy_2Fe_{14}B$	8.76	12.01	14.0	0.71	~ 15	598

The anisotropy field H_a for $Dy_2Fe_{14}B$ and $Tb_2Fe_{14}B$ is two, and respectively three times higher comparing to $Nd_2Fe_{14}B$. For the heavy rare-earth elements (HREE, classified by atomic number $Z > 62$) the rare-earth sublattice moment couples antiferromagnetically with the Fe sublattice moment whereas for the case of light rare-earth elements (LREE, atomic number Z from 57 to 61) the coupling is ferromagnetic. Therefore Dy and Tb substituted for Nd in $Nd_2Fe_{14}B$ significantly enhance anisotropy but this comes at the expense of saturation magnetization and limited amounts of these elements will be used in alloy formulations.

In the case of the $Ce_2Fe_{14}B$ compound, the anisotropy field H_a is only a third of the value corresponding to $Nd_2Fe_{14}B$. The lattice constants of $Ce_2Fe_{14}B$ are significantly smaller comparing to $Pr_2Fe_{14}B$ and $Nd_2Fe_{14}B$, making a noticeable exception in the lanthanide contraction trend illustrated on $R_2Fe_{14}B$ compounds (Table 2.1). This is an indication that Ce is tetravalent in $Ce_2Fe_{14}B$. Tetravalent Ce does not carry magnetic moment (nor does La) so the anisotropy of the compound is supported mainly by the exchange interactions in the Fe sublattice [34, 35].

The Curie temperatures of the $R_2Fe_{14}B$ compounds are determined by the exchange coupling interactions between the atoms in the crystal lattice (Fe-Fe - the strongest, R-Fe and R-R) which depend on the electronic configuration of the rare-earth and transition metal components and are sensitive to lattice spacing. The distances between near neighbor Fe atoms differ according to their position on the lattice sites. Based on the Bethe-Slater curve model, for Fe-Fe distances lower than $\approx$ 2.50 Å (the nearest neighbor distance in α-Fe), the exchange interaction energy values are negative (antiferromagnetic coupling) while at larger distances the interaction energy is positive (ferromagnetic coupling), the negative exchange interactions contributing to the significantly lower Curie temperatures of these compounds comparing to pure Fe (1043 K).

The ordering temperature can be raised by substituting Fe atoms and/or increasing the unit cell volume [36, 37]. Partially substituting Co on the Fe sites favors the positive interactions (exchange energies $J_{Co\text{-}Co}$ and $J_{Co\text{-}Fe}$ are higher than $J_{Fe\text{-}Fe}$) which results in the monotonous increase of T_C with x in $R_2Fe_{14\text{-}x}Co_xB$ compounds.

For instance in $Nd_2Fe_{14-x}Co_xB$ it increases from 585 K at x = 0 to 700 K at x = 2, reaching 1000 K for $Nd_2Co_{14}B$. However the Co sublattice favors basal plane magnetization decreasing anisotropy and for $x \geq 9$ a uniaxial to basal plane spin reorientation occurs below T_C limiting the practical use of Co in this range [38].

The interstitial insertion of hydrogen atoms by hydrogenation induces lattice expansion without altering the crystal symmetry and results in a higher T_C for $R_2Fe_{14}BH_x$ hydrides comparing to the parent compound, the phenomenon being ascribed to the increase in Fe-Fe distances and related reduction of negative exchange interactions [39, 40]. Other studies imply that relating the T_C increase upon hydrogenation only to the increased distances in the Fe sublattice is an oversimplification and show that rare-earth sublattice also plays a major role (particularly for $HRE_2Fe_{14}B$) because the presence of hydrogen influences the electronic structure of the entire system [41]. Hydrogenation of $Nd_2Fe_{14}B$ increases the saturation magnetization M_s but decreases significantly the anisotropy field (H_a) [42] making the hydride unusable as permanent magnet but it is employed in alloy processing being an effective powder production method (Hydrogen decrepitation, detailed in Section 2.3.3).

2.3 $Nd_2Fe_{14}B$-based alloys

2.3.1 Phase equilibria and solidification behavior

All metallurgical processes, from synthesis to casting, forming (such as hot-pressing and hot-deformation) and heat treatments (like annealing for various purposes or sintering) are designed according to the solidification and phase formation behavior of the targeted alloy which is described by the phase diagram of the respective alloy system.

Depending on composition and cooling conditions, $Nd_2Fe_{14}B$-based permanent magnet alloys solidify in a complex multiphase configuration consisting of the main $Nd_2Fe_{14}B$ hard magnetic phase (Φ-phase) and other non-magnetic or soft magnetic phases. The isothermal section at 1000 °C of the Fe-rich corner in Nd-Fe-B phase diagram is introduced in Figure 2.9.

Three equilibrium phases are predicted by the ternary phase diagram: the tetragonal $Nd_2Fe_{14}B$ (Φ-phase), the tetragonal boride $Nd_{1+\varepsilon}Fe_4B_4$ (η-phase) and the low-melting-point Nd-rich phase. In the Nd-Fe-B phase diagram, the phase of interest, $Nd_2Fe_{14}B$ (for the stoichiometric composition $Nd_{11.76}Fe_{82.35}B_{5.88}$ in at.% or $Nd_{26.68}Fe_{72.32}B_{1.00}$ in wt. %) is formed from the melt by peritectic reaction of the liquid with Fe at 1180 °C (Figure 2.10). The primary crystallization product is Fe as the peritectic Φ-phase melts incongruently.

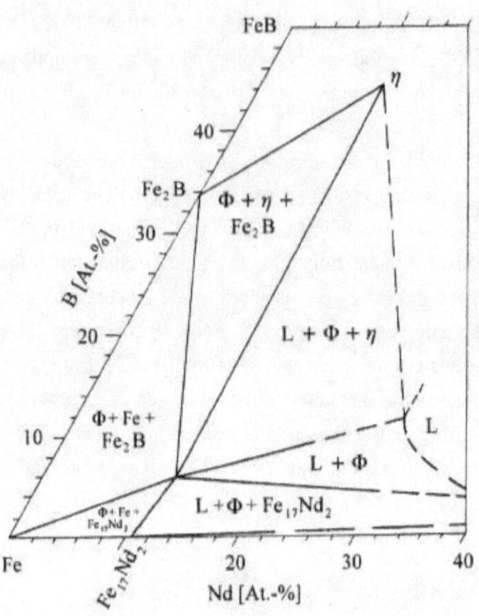

Figure 2.9: Isothermal section at 1000 °C in the Fe-rich corner of the Nd-Fe-B system [43].

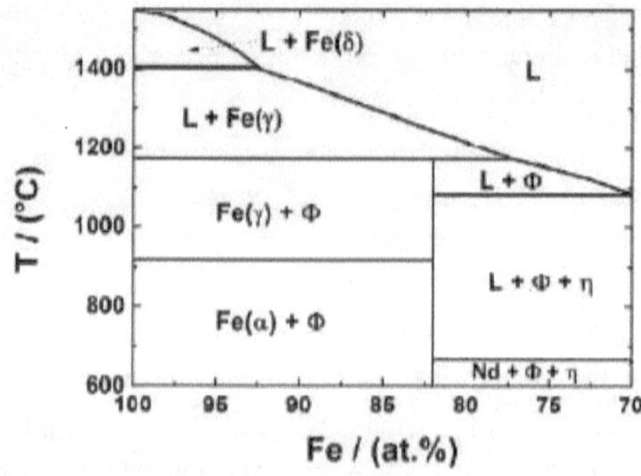

Figure 2.10: Pseudo-binary phase diagram with fixed Nd:B = 2:1 ratio [44].

The solidification of over-stoichiometric alloys (above 12 at. % Nd) ends with the eutectic formation of the Nd-rich phase at 665 °C [45]. For ≥15 at. % Nd, alloys having < 6 at. % B are composed of the Φ-phase and Nd+Fe in the as-cast structure while those having > 6 at. % B show the η-phase but no Fe [46]. Both Fe and η-phase segregations have a detrimental effect on the magnetic properties. Iron has a negative impact on coercivity.

The η-phase formation reduces the proportion of Nd$_2$Fe$_{14}$B hard magnetic phase with a negative impact on remanence. For rare-earth lean alloys with slightly over-stoichiometric composition (enough to form the Nd-rich phase), the η-phase formation is avoided the segregation of α-Fe through peritectic reaction is suppressed by rapid quenching of the melt.

The formation of the Nd-rich intergranular phase was identified as an essential factor in the development of high coercivity Nd$_2$Fe$_{14}$B-based permanent magnets [47]. The Nd-rich intergranular phase is a non-magnetic spacer that separates and magnetically decouples the hard magnetic grains giving rise to coercivity by impeding the nucleation and growth of reverse domains.

2.3.2 Fast solidification techniques

Melt-spinning

Melt-spinning is a metallurgical method that allows the production of amorphous or partially amorphous nanocrystalline alloys in the form of thin ribbons. It is a fast solidification technique that can achieve extremely high cooling rates in the order of 10^5 - 10^6 K/sec [48, 49]. The process is conducted in inert gas (Ar or He) atmosphere. It involves the induction melting of the alloy in a crucible having a nozzle at the bottom and ejecting the melt through the nozzle by means of gas pressure onto a water chilled rotating wheel (Figure 2.11). The crucible is most commonly made of quartz or boron nitride (BN) and the nozzle geometry can be circular or in the shape of a rectangular slit.

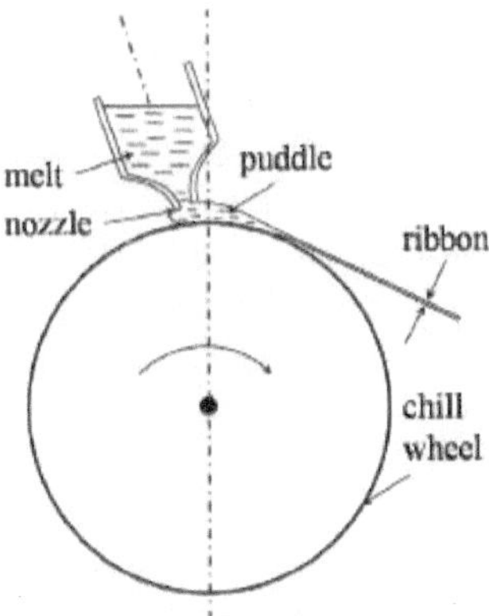

Figure 2.11: Schematic illustration of the ribbon formation at the contact region between the molten material and the rotating wheel, redrawn and adapted from ref. [49].

The $Nd_2Fe_{14}B$ melt-spun material typically takes the shape of brittle ribbons/flakes of varying lengths on the millimeter or centimeter scale, with thicknesses of 30 - 50 μm and widths of 1 - 3 mm. The cooling rate of the melt onto the quenching wheel determines the melt-spun alloys microstructure.

The technique allows the control of the cooling rate mainly through the variation of the wheel speed. The grain size, the grain size distribution and the alloys phase composition have a decisive impact on its magnetic properties. A uniform microstructure composed of $Nd_2Fe_{14}B$ grains with diameters of ~30 nm completely surrounded by a thin (~ 2 nm) Nd-rich phase was directly related to maximized magnetic properties in $Nd_2Fe_{14}B$-based melt-spun alloys. The quench window that yields the optimum microstructure is narrow (in the range of a few m/s). It is therefore difficult to achieve the optimum microstructure in a single step consisting of the direct quenching from the melt. The usual practice involves over-quenching to amorphous or partially amorphous state and annealing the melt-spun material to achieve uniform crystallization and optimal microstructure [50]. For best results a steady melt flow and a high and uniform heat extraction rate are required so, besides the wheel speed, other melt-spinning processing factors that impact the cooling conditions (therefore microstructure homogeneity) are the quenching gas and the wheel material (keeping in mind their heat conductivities), the gas chamber pressure and ejection pressure, the wheel surface roughness, system vibrations etc. [51].

The melt-spun material is magnetically isotropic as it is composed of randomly oriented $Nd_2Fe_{14}B$ grains. The melt-spun $Nd_2Fe_{14}B$ material is mainly used for the fabrication of *isotropic polymer-bonded magnets* and *anisotropic hot-deformed magnets* [38]. *Anisotropic polymer-bonded magnets* can also be prepared from decrepitated and milled hot-deformed material [52]. *Isotropic bonded magnets* are composed of melt-spun powder consolidated into the specific shape using epoxy resin (by compression molding - 2 wt. % binder content) or thermoplastic polymers (by injection molding - 8 to 15 wt. % binder content) [53].

Full density is achieved by consolidating the powders through uniaxial *hot-pressing* and *hot-deformation (die-upsetting)*, the latter also inducing structural and magnetic anisotropy. The hot-pressed body is magnetically isotropic but through plastic deformation of the grains alignment parallel to the *c*-axis direction (or texture) is induced. For deformation the hot-pressed body is placed in a die cavity of a larger diameter which allows the lateral plastic flow of the material resulting in height reduction of the compact and grain elongation perpendicular to the press direction. The resulting deformed microstructure features flat, platelet-like shaped grains. The hot-pressing and hot-deformation temperatures are typically above the melting point of the intergranular Nd-rich phase to facilitate densification through grain flow/rotation and plastic deformation in the latter stage. The alloys deformation behavior was associated with the solution-precipitation creep model [54].

The presence of the liquid phase is not mandatory for a certain degree of deformation to take place in the $Nd_2Fe_{14}B$ grains (plastic deformability is determined by the crystal lattice characteristics - number of slip systems, dislocation mobility) but it is necessary in the process in order to avoid cracking.

The liquid phase assistance is therefore of technological importance because it enhances the alloys hot-workability. Hot-workability can be tuned by the addition of alloying elements that decrease the melting point of the rare-earth-rich phase such as Cu [55] or Ga and Co (as in the commercial Magnequench MQU-F grade melt-spun alloy ($Nd_{13.6}Fe_{73.6}Co_{6.6}Ga_{0.6}B_{5.6}$ at. %)) [56]. Optimized hot-workability allows for strong texture to be achieved which determines the rise in magnetization in hot-deformed anisotropic magnets compared to the isotropic hot-pressed ones.

Strip-casting

Strip-casting is also a fast solidification metallurgical process working on a principle similar to melt-spinning. It involves induction melting the alloy and pouring it on a chilled rotating wheel through a ceramic tundish system. The cooling rates achieved by this process are in the range of 10^3 - 10^4 K/s. The strip-cast $Nd_2Fe_{14}B$ material takes the shape of several centimeters wide flakes with thicknesses of roughly 250 - 350 μm. The microstructure of the strip-cast material features $Nd_2Fe_{14}B$ grains with plate-like shape surrounded by a thinly distributed Nd-rich phase (60 - 150 nm thick). The grains appear lamellar on the flake cross-section, oriented parallel to the heat-extraction direction, 5 - 25 μm thick toward the wheel side and 25 - 60 μm thick close to the free surface. Comparing to conventional ingot casting, due to the high solidification rate, strip-casting produces a homogenous microstructure without large accumulations of Nd-rich phase at the grain boundaries and free of α-Fe dendrites segregations [57]. An energy product $|BH|_{max}$ as high as 460 kJ/m^3 (~96 % of the theoretical value for $Nd_2Fe_{14}B$) was attained in anisotropic sintered magnets using optimized strip-cast material with a nearly-stoichiometric rare-earth content [58, 59].

2.3.3 Hydrogen processing of $Nd_2Fe_{14}B$ alloys

Hydrogen decrepitation (HD)

The HD treatment of $SmCo_5$ and Sm_2Co_{17} alloys was patented by Harris *et al.* [60] in 1979 and it was later applied to the $Nd_2Fe_{14}B$ system. For $Nd_2Fe_{14}B$ alloys, the process involves the room temperature exposure of the cast material to hydrogen gas in a an evacuated and sealed reaction vessel at slight overpressure. Hydrogen is absorbed along the grain boundaries being diffused into the Nd-rich intergranular phase and then into the matrix phase. Hydrogen absorption is dissociative in these materials, molecular hydrogen being dissociated into hydrogen atoms which will occupy interstitial sites. The interstitial absorption with the exothermic formation of NdH_{2+x} further fuels the hydrogenation of the matrix to form $Nd_2Fe_{14}BH_x$ (Figure 2.12).

The presence of the Nd-rich phase therefore plays a crucial role absorbing hydrogen readily at room temperature while the $Nd_2Fe_{14}B$ single phase material absorbs hydrogen at around 160 °C. Decrepitation occurs due to the large difference in volume expansion upon hydrogenation between the $Nd_2Fe_{14}B$ phase and the Nd-rich phase (volume increase of 4.8 % and 16.4 % respectively) and the material is thus turned into friable powder [61, 62, 63].

The material fractures along the grain boundaries in a first stage and then microfractures are produced through the grains [64]. Within the polycrystalline grain, hydrogen diffuses more easily along the defects like subgrain boundaries (or low angle grain boundaries) thus producing intragranular microcracking. The brittle coarse powder resulted after HD is milled and further used for the fabrication of anisotropic magnets by sintering which is performed in vacuum to facilitate hydrogen thermal desorption. HD is an useful, economical step employed for the pulverization of strip-cast or ingot material in the production of $Nd_2Fe_{14}B$ permanent magnets by powder metallurgy route.

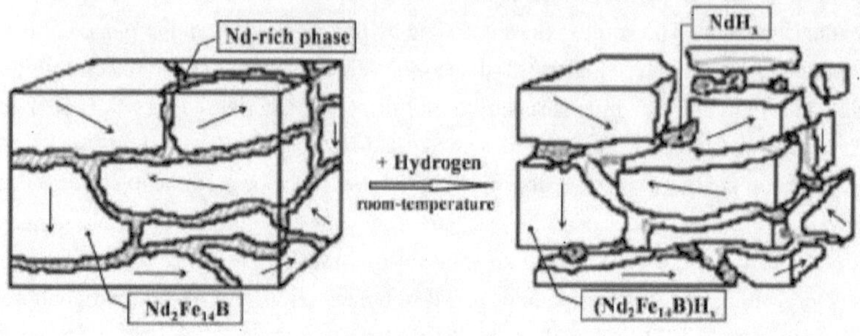

Figure 2.12: Schematic illustration of hydrogen decrepitation (HD) process in $Nd_2Fe_{14}B$ alloys.

Dehydrogenation, achieved through the vacuum annealing of the alloy, occurs in three stages of thermal desorption, from the matrix phase at 200 °C and then gradually from the rare-earth-rich phase, in two transformation sequences ($NdH_3 \rightarrow NdH_2$ peaks at 250 °C and $NdH_2 \rightarrow Nd$ at 600 °C) [65]. Calorimetric data on the vacuum desorption behavior (onset, peak and offset temperatures) of some rare-earth binary hydrides can be found in V. A. Yartys et al. 1997 [66]. For instance $CeH_3 \rightarrow CeH_2$ onsets at 150 °C, peaks at 340 °C and is completed at 400 °C and $CeH_2 \rightarrow Ce$ onsets at 500 °C, peaks at 650 °C and is completed at 730 °C, significantly lower temperatures comparing to the Nd counterpart ($NdH_3 \rightarrow NdH_2$ peaks 380 °C while $NdH_2 \rightarrow Nd$ peaks 720 °C). The $CeFe_2$ Laves compound, known to segregate at the grain boundaries in Ce-substituted $Nd_2Fe_{14}B$ and $Ce_2Fe_{14}B$ alloys (details in Section 2.4.1), exothermically absorbs hydrogen at room temperature and transforms simultaneously into amorphous $CeFe_2H_x$ which decomposes (if further heated, at temperatures above ~270 °C) into CeH_2 and α-Fe [67]. The more recent calorimetric study by Dilixiati et al. [68] suggests that upon heating in hydrogen atmosphere, the amorphous $CeFe_2H_x$ phase undergoes a two step decomposition reaction, first into $Ce_{1-y}Fe_2H_x$ and CeH_2 and finally into CeH_2 and α-Fe.

The hydrogen absorption-desorption behavior is therefore influenced by the composition and the microstructure of the cast alloy. Besides the composition of the matrix phase, the condition of the intergranular material (proportion, thickness, distribution, chemical composition, structural state, phase composition, presence of segregated phases) as critical diffusion pathway in the process requires analysis and control if room temperature decrepitation and, furthermore, optimal milling and sinterability properties are aimed for. The hydrogen absorption-desorption characteristics have a decisive impact on the response of the material to the HDDR treatment and on the decrepitated powder sintering and need to be considered according to the adopted processing route.

Hydrogenation - disproportionation - desorption - recombination (HDDR)

The HDDR process refers to a sequence of H_2 pressure and temperature dependent transformations experienced by some hydrogen-absorbing intermetallic compounds. This phenomenon was observed for both $SmCo_5$ and $Nd_2Fe_{14}B$ permanent magnet alloys during sintering experiments in hydrogen atmosphere [69, 70].

In the case of $Nd_2Fe_{14}B$ alloys a typical HDDR treatment involves heating the material in H_2 atmosphere, the process entailing four reaction stages, with the following transformations:

(i) *hydrogenation* - hydrogen is absorbed and the $Nd_2Fe_{14}BH_x$ phase is formed;

(ii) *disproportionation* - $Nd_2Fe_{14}BH_x$ decomposes into $NdH_{2\pm x}$, Fe_2B and α-Fe at temperatures above 600 °C;

(iii) *desorption* - process temperature is raised and the system is evacuated so that hydrogen is desorbed from $NdH_{2\pm x}$;

(iv) *recombination* - quasi-simultaneously with hydrogen desorption, the resultant highly reactive Nd recombines with Fe_2B and α-Fe into the original $Nd_2Fe_{14}B$.

The HDDR reversible reactions sequence [71] can be written as follows:

$$Nd_2Fe_{14}B + \frac{1}{2}xH_2 \leftrightarrow Nd_2Fe_{14}BH_x \tag{2.10}$$

$$Nd_2Fe_{14}B + (2 \pm x)H_2 \leftrightarrow 2NdH_{2\pm x} + 12\alpha - Fe + Fe_2B \pm \Delta H \tag{2.11}$$

Throughout the different reaction stages, the HDDR process induces substantial transformations in the alloy microstructure (documented in detail in Figure 2.13). The HDDR process is locally confined taking place within the frame of the initial cast grain structure. The finely divided mixture of $NdH_{2\pm x}$, α-Fe and Fe_2B resulted from disproportionation recombines into sub-micron sized $Nd_2Fe_{14}B$ grains (down to ~300 nm, close to the dimension of the $Nd_2Fe_{14}B$ magnetic monodomain) upon hydrogen desorption [71, 72].

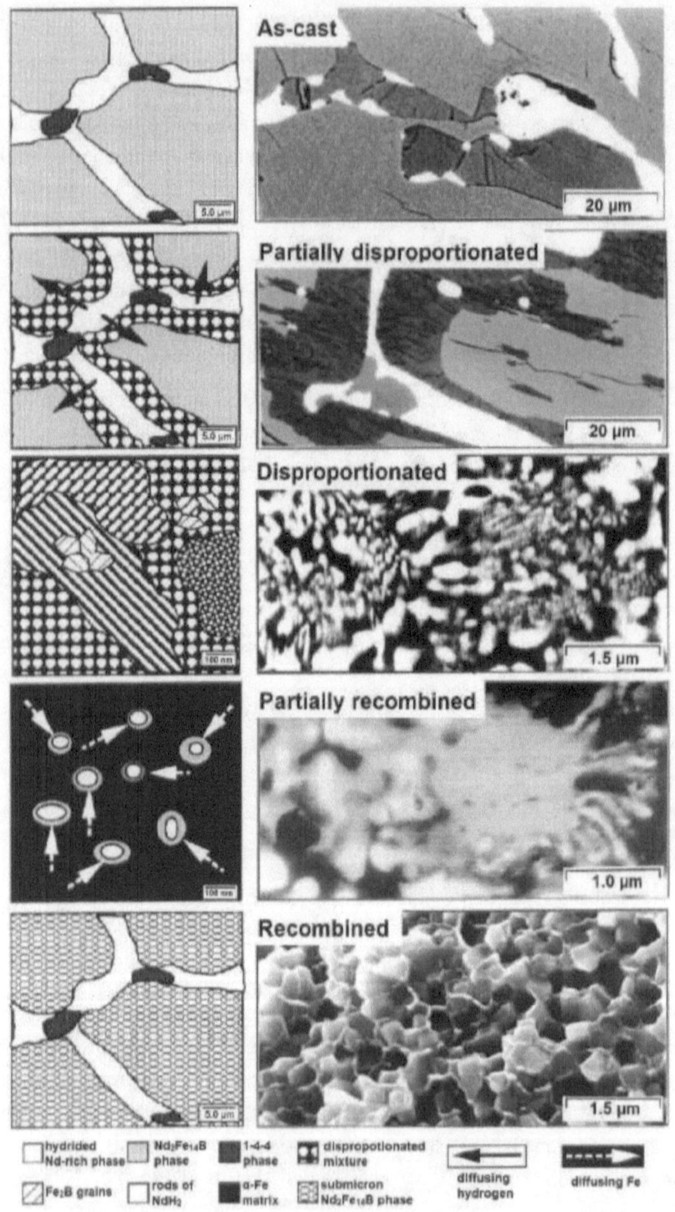

Figure 2.13: Schematic illustrations/SEM micrographs of the microstructural transformations induced in a Nd$_2$Fe$_{14}$B-based alloy (composed of Φ, η and Nd-rich phases in as-cast state) during HDDR process [71].

Apart from the grain size reduction, the HDDR treatment also induces texture. It allows the direct production of highly anisotropic magnets through powder alignment and consolidation.

The crystallographic orientation relationships between the parent, disproportionated and recombined phases suggest that Fe_2B is the anisotropy mediating phase throughout the HDDR process stages. The Fe_2B tetragonal structure retains the initial orientation of the $Nd_2Fe_{14}B$ crystallite and further acts like a "backbone" for the $Nd_2Fe_{14}B$ phase to recombine during HDDR, the concept of "texture memory effect" (TME) being coined for the description of this phenomenon [73, 74, 75, 76]. Earlier studies have suggested that additives like Zr, Nb, Ga, Co play a role in the texture development in HDDR alloys [77, 78, 79] due to their tendency to form borides, this presumably adding to the stability of the Φ boride phase during the treatment thus allowing the retention of its initial alignment. However, it was demonstrated that texture inducement is independent of the alloys additive content and highly anisotropic powders can be produced when the rates of disproportionation and recombination reactions are rigorously controlled through optimized parameters (H_2 pressure and desorption rate, annealing temperatures and dwell times) in a dynamic HDDR (d-HDDR) process [80, 81].

The disproportionated microstructure exhibits $NdH_{2\pm x}$ and α-Fe in a fine lamellar-shaped structure at the initial disproportionation reaction stages and changes to granular morphology in advanced stages [72]. A recent study suggested that lamellar NdH_x - α-Fe disproportionation structures can be correlated to high magnetic anisotropy in d-HDDR powders [82]. It claims that the lamellar regions within the disproportionated grain recombine into coarser $Nd_2Fe_{14}B$ grains while crystallographic alignment coherency is maintained through desorption-recombination reaction which results in textured d-HDDR powders. However, the literature refutes the link between the lamellar disproportionated microstructure and texture. It is generally accepted that texture inducement is mediated by the tetragonal Fe_2B phase and is influenced by the hydrogen pressure and desorption rate during absorption and desorption, being achieved regardless of the degree of spheroidization.

The HDDR powders are generally destined for resin blending and forming into anisotropic bonded magnets [83, 31]. Due to the reduced sintering temperature and time (thus avoiding grain growth as opposed to the conventional sintering route) spark plasma sintering (SPS) was also used for the HDDR powder consolidation resulting in good magnetic properties (with an optimized $|BH|_{max} = 210$ kJ/m^3) [84].

2.4   Cerium/Lanthanum substituted $R_2Fe_{14}B$-based permanent magnets

2.4.1   Phase structure and microstructure

The $Ce_2Fe_{14}B$-based alloys have a specific solidification behavior, different from $Nd_2Fe_{14}B$-based ones due to the formation of $CeFe_2$ as intergranular phase instead of the rare-earth-rich phase (Figure 2.14). The tetragonal 2:14:1 hard magnetic phase (Φ-phase) and the η-borides ($Nd_{1+\varepsilon}Fe_4B_4$ and $CeFe_4B_4$ phases) are analogous between the two systems.

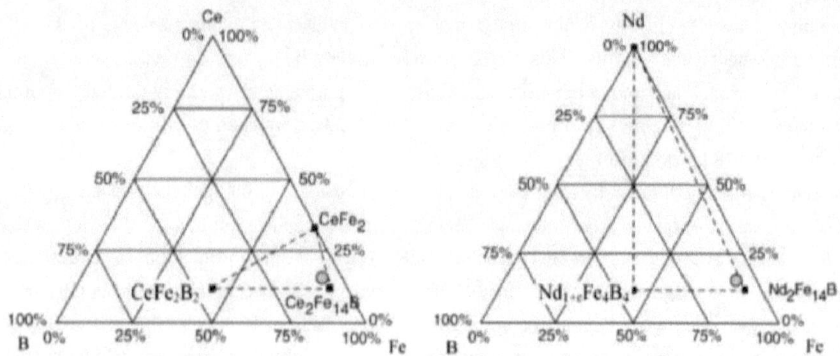

Figure 2.14: Phase equilibria in Ce-Fe-B and Nd-Fe-B systems at 25 °C [21].

$CeFe_2$ forms by peritectic reaction and is a paramagnetic phase of $MgCu_2$-type cubic symmetry [85]. Similarly to the Nd-rich phase [86], $CeFe_2$ can act as a magnetically decoupling spacer between the Φ-phase grains thus playing a important role in coercivity development in $Ce_2Fe_{14}B$-based magnets [87]. $CeFe_2$ does however have a higher melting point (starts melting at 925 °C) comparing to the Nd-rich phase (655 °C) and also shows a tendency to aggregate at the triple-point regions between the matrix phase grains which results in an inferior grain boundary wetting ability during sintering. In these conditions, sintering above the melting point of $CeFe_2$ would allow densification and grain decoupling through its distribution along the matrix phase grain boundaries which results in a continuous intergranular phase.

In $Ce_2Fe_{14}B$ alloys and Ce-substituted $Nd_2Fe_{14}B$ alloys the solidified phase structure will depend of course on the Ce content and the cooling rate. Above the stoichiometric composition in $Ce_2Fe_{14}B$-based alloys, with increasing Ce concentration, the $CeFe_2$ phase increases in proportion. The increasing proportion of $CeFe_2$ comes at at the expense of other possible minor phases like Fe_3B, $CeFe_7$ as shown for melt-spun $Ce_{12+x}Fe_{82-x}B_6$ (x = 0, 0.5, 1, 1.5, 2 … 3) [88].

In the case of melt-spun alloys, Ce-rich overstoichiometric compositions were shown to yield best magnetic properties. For instance, the $Ce_{17}Fe_{78}B_6$ melt-spun alloy was found to be the optimum in Ce-Fe-B system, composition which is significantly richer in rare-earth component comparing to the $Nd_{13}Fe_{82}B_5$ optimum for Nd-Fe-B system [89].

For substituted $(Nd_{1-x}Ce_x)_2Fe_{14}B$ alloys, it was reported by K. A. Gschneidner *et al.* [18] and A. K. Pathak *et al.* [90] that the substitutional solid solution range extends from x = 0 to x ≈ 0.15 and from x ≈ 0.4 to x = 1 concentrations, with a state of phase separation due to restricted solubility in the $Ce_2Fe_{14}B$ - $Nd_2Fe_{14}B$ system in x ≈ 0.15 ... ≈ 0.4 range.

The state of phase separation was associated with an anomalous increase in lattice constants at x ≈ 0.25 Ce (most likely due to the mixed-valent behaviour of Ce which involves its atomic radius variation, hence change in solid solubility according to the empirical alloying theory - Hume-Rothery rule - and the lattice constants deviation from linearity - implied by Vegard's law).

In the same studies, the above-mentioned separation state (evidenced later by HRTEM [91]) was also related to an increase in coercivity in x = 0.2 ... 0.3 (a deviation from the decreasing trend with increasing Ce concentration) and a depression of the demagnetization loop squareness which was found to restore in alloys at x > 0.3. At Ce fractions below 25 % of the rare-earth content, Ce partly substitutes Nd in the intergranular material forming the (Ce, Nd)-rich phase (with a lower melting point comparing to the Nd-rich phase [92]). At higher concentrations, $CeFe_2$ segregations occur in substituted magnets [87]. Above 50 % Ce concentration $CeFe_2$ replaces the (Ce, Nd)-rich phase in significant proportion and additional α-Fe and Ce_2Fe_{17} segregations can also occur at Ce concentrations higher than 80 % [21]. Depending on the Ce concentration in the alloys, the intergranular material in Ce-substituted magnets consists of a (Ce, Nd)-rich phase and/or $CeFe_2$. Unlike the rare-earth-rich intergranular phase which tends to solidify uniformly around the Φ phase, in $Ce_2Fe_{14}B$ alloys $CeFe_2$ is rather discontinuous and tends to aggregate at the triple points. The non-uniform segregation of $CeFe_2$ intergranular phase alters the lamellar structure of the Φ-phase grains in Ce-Fe-B strip-cast alloys [93]. It was shown that $CeFe_2$ segregations in $Ce_2Fe_{14}B$ alloys can be significantly reduced or eliminated through La substitution for Ce while a more uniformly distributed La-Ce intergranular phase is formed instead. The effect was shown in (La, Ce)-Fe-B strip-cast [94]. and in melt-spun alloys [95] where La substitution also induced microstructure refinement reducing the grain size [96]. Cooling rate also is here important as $CeFe_2$ formation is suppressed at lower La fractions for melt-spun alloys comparing to strip-cast ones. Also the substitution of Fe by Si was proved to restrict the formation of $CeFe_2$ reducing its volume fraction in $Ce_{17}Fe_{78-x}Si_xB_6$ melt-spun ribbons at x ≥ 0.75 concentrations [97]. The $La_2Fe_{14}B$ compound is unstable, difficult to obtain, being prone to strong segregations of α-Fe and La-B in both as-cast and melt-spun form [98]. Having a larger atomic radius, La has limited solubility in the $Nd_2Fe_{14}B$ phase so it is expected to be expelled to the grain boundaries [99, 100]. However, in doping quantities, La was shown to reduce the grain size in $Nd_2Fe_{14}B$-based melt-spun alloys [101] (it is an effective glass former used to obtain amorphous alloys through rapid quenching [102]).

2.4.2 Approaches for hard magnetic properties enhancement

Substituting Nd by Ce or/and La results in decreased performance of magnets since $Ce_2Fe_{14}B$ and $La_2Fe_{14}B$ compounds have lower intrinsic magnetic properties compared to $Nd_2Fe_{14}B$ (detailed in section 2.2). However, considering the significant cost reduction it involves, the use of Ce and La is attractive because it allows the fabrication of "gap magnets", hard magnets of intermediate performance ($|BH|_{max}$ of 100 - 200 kJ/m^3) between hexaferrites (< 38 kJ/m^3) and $Nd_2Fe_{14}B$-based magnets (> 200 kJ/m3) [103]. This ultimately involves the replacement of the critical rare-earth component up to a level where the maximum acceptable dilution of magnetic properties in $(Nd_{1-x}R_x)_2Fe_{14}B$ alloys is reached or the magnetic hardening of $Ce_2Fe_{14}B$-based alloys. Microstructure engineering through composition (relative rare-earth content and alloying elements), solidification and thermal treatment optimization is of critical importance to maximize coercivity and remanence. In addition, the use of alloying elements that modify the coordination environment of the Ce atom site increasing its steric volume is thought to shift the Ce valence toward magnetic moment carrying Ce^{3+} state.

Alloy composition optimization and microstructure tuning

First studies aiming for the development of inexpensive RE-Fe-B permanent magnets date back to the mid 80s. M. Okada et al. [104] used Ce-dydimium in $RE_2Fe_{14}B$ sintered magnets (dydimium is a Nd-Pr mixture, an intermediate product in the process of Nd separation). It was then observed that Ce improves densification during sintering as it decreases the melting point of the rare-earth-rich grain boundary phase and of the matrix phase. Also, for Ce and Ce-dydimium containing sintered magnets it was later shown that additives like Si, Dy_2O_3 and Co can partially compensate for the loss in coercivity and respectively Curie temperature caused by Ce substitution [105, 106]. Alternatively, mischmetal (50% Ce, 25% La, 15% Nd, 5% Pr, others 5%, a by-product mixture resulting from rare-earths separation process) was also introduced as cheap substitute for Nd by Gong and Hadjipanayis (1988) who used it to fabricate melt-spun ribbons [107]. A 1991 study reported the fabrication of Ce substituted sintered magnets of composition $Ce_{8.1}Nd_{5.4}Fe_{62}Co_{17}SiB_{6.5}$ at. % with $B_r = 11.7$ kG (1.17 T), $H_{cj} = 7.5$ kOe ($\mu_0H_c = 0.75$ T) and $|BH|_{max} = 27.2$ MGOe (216.45 kJ/m^3) compared to the $Ce_{13.5}Fe_{62}Co_{17}SiB_{6.5}$ at. % magnet that measured a $|BH|_{max} = 11.1$ MGOe (88.33 kJ/m^3) and also highlighted the positive effect of Si on coercivity (the addition of 1 at. % doubles the coercivity while keeping T_C rather unchanged while it partially substitutes for Fe) [106]. In another early study [108], improved densification/sinterability and good magnetic properties due to Ce and Al partial substitution were observed in La containing sintered magnets of composition $[Nd_{0.6}(La_{0.33}Ce_{0.67})_{0.4}]_{15.5}(Fe_{0.96}Al_{0.04})_{77}B_{7.5}$ ($B_r = 1.02$ T, $\mu_0H_c = 0.81$ T, $|BH|_{max} = 167.1$ kJ/m^3).

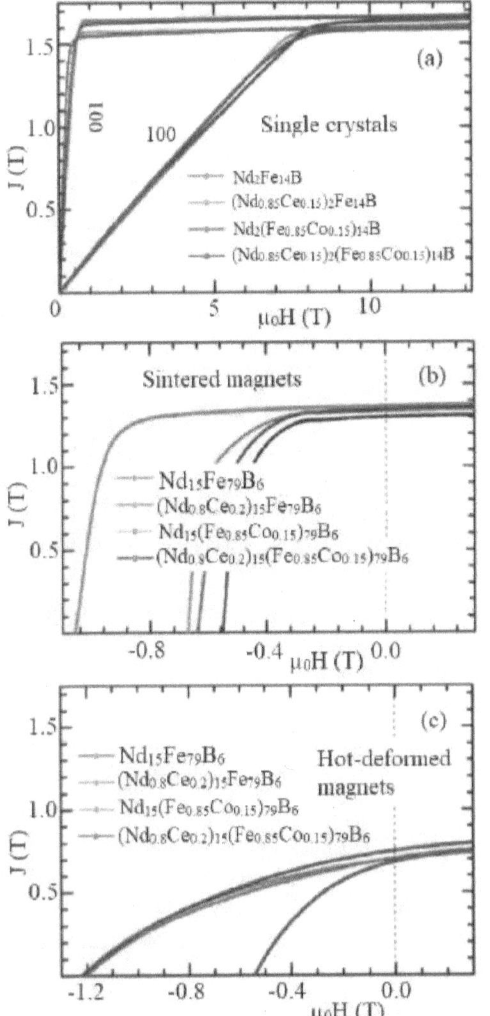

Systematic efforts have been focused particularly on the research and development of critical rare-earth lean and critical rare-earth free RE$_2$Fe$_{14}$B permanent magnet alloys since 2012. From the assessment of the intrinsic magnetic properties of Ce substituted Nd$_2$Fe$_{14}$B single crystals to the evaluation of alternatives for alloying and processing, solutions were sought for with the aim to minimize or, ideally, completely eliminate the heavy rare-earth additives and the critical rare-earth component in RE$_2$Fe$_{14}$B magnets. (Nd$_{1-x}$Ce$_x$)Fe$_{14}$B single crystals with up to 38 % Ce retained 97.8 % of the saturation magnetization of the pure compound and measured an anisotropy field μ_0H$_A$ = 5.2 T [109].

Figure 2.15: Magnetization vs. field curves along the easy and hard axis for (Nd$_{1-x}$Ce$_x$)$_2$(Fe$_{1-y}$Co$_y$)$_{14}$B single crystals, x = 0, 0.15; y = 0, 0.15 (a); (Nd$_{1-x}$Ce$_x$)$_{15}$(Fe$_{1-y}$Co$_y$)$_{79}$B$_6$, x = 0, 0.2; y = 0, 0.15 sintered (b) and hot-deformed magnets (c), redrawn from ref. [110].

Measurements on single crystals of (Nd$_{1-x}$Ce$_x$)$_2$(Fe$_{1-y}$Co$_y$)$_{14}$B formula with x = 0, 0.15 and y = 0, 0.15 (Figure 2.15 a) indicate that a small fraction of Ce does not bring about a significant decrease in saturation magnetization and the anisotropy is only mildly affected (as is the case for the Ce and Co co-substitutued crystal) while 0.15 Co fraction has no impact on saturation magnetization and anisotropy [110].

Microstructural factors such as phase segregations and grain size [111] which cause the decreasing coercivity in Ce-substituted sintered magnets (as shown in Figure 2.15 b) can be ameliorated through alloy processing by rapid solidification which restricts the formation of $CeFe_2$ and reduces the grain size (hot-deformed magnets fabricated from melt-spun alloys Figure 2.15 c) thus enhancing coercivity. Magnetic properties decrease with increasing Ce fractions in Nd-Fe-B melt-spun alloys [112]. At 20 % Ce replacement, coercivity H_{ci} ($\mu_0 H_c$) of 17.7 (1.77) , 13.7 (1.37) and 9.4 (0.94) kOe (T) and $|BH|_{max}$ of 12.6 (100.3), 12.8 (101.9) and 31.2 (248.3) MGOe (kJ/m^3) were registered for $(Nd_{0.8}Ce_{0.2})_{2.4}Fe_{12}Co_2B$ alloy in isotropic melt-spun, hot-pressed and respectively hot-deformed state while Zr and C added in this alloy to form ZrC impedes grain growth and contributes to further enhancement of the melt-spun alloy magnetic property (up to H_{ci} = 18.9 kOe ($\mu_0 H_c$ = 1.89 T) and $|BH|_{max}$ = 15 MGOe (119.37 kJ/m^3)) through microstructure refinement [90]. As a result of Φ phase separation within $0.2 \leq x \leq 0.3$ an abnormal increase in coercivity was observed at about 0.25 Ce fraction [113].

There are few studies on La individually substituted in $Nd_2Fe_{14}B$ magnets and most deal with α-$Fe/Nd_2Fe_{14}B$ nanocomposite alloys which have substoichiometric compositions. Only recently, an energy product $|BH|_{max}$ = 100 kJ/m^3 was reported for $(Nd_{1-x}La_x)_{13}Fe_{81}B_6$ melt-spun ribbons with x = 0.4 La concentration, higher comparing to x = 0.3 due to refined grain structure and La segregated at the grain boundaries [114].

Also mischmetal containing $RE_2Fe_{14}B$ alloys [115, 116] have again caught attention due to their reduced material cost and environmental impact.

For $Ce_2Fe_{14}B$-based magnets, the rare-earth content and microstructure optimization proved to be decisive in magnetic property development. In melt-spun alloys, Ce-rich, overstoichiometric compositions were shown to give best results. For instance, coercivity increased monotonically (up to a maximum of 6.54 kOe ($\mu_0 H_c$ = 0.654 T)) with increasing Ce concentration and $CeFe_2$ intergranular phase fraction in $Ce_{12+x}Fe_{82-x}B_6$ (x = 0, 0.5, 1, 1.5, 2 and 3) [88]. From a large range of tested formulations, J. F. Herbst et al. 2012 [89] reported $Ce_{17}Fe_{78}B_6$ melt-spun alloys as the composition giving optimal magnetic properties (B_r = 4.9 kG (0.49 T), H_{ci} = 6.2 kOe ($\mu_0 H_c$ = 0.62 T), $|BH|_{max}$ = 4.1 MGOe (32.6 kJ/m^3)). Cobalt substitution for Fe in $Ce_{17}Fe_{78}B_6$ ($Ce_3Fe_{14}B$) alloy increased the Curie temperature to 516 K for $Ce_3Fe_{12}Co_2B$ composition at slight decrease of coercivity (B_r = 5.2 kG (0.52 T), H_{ci} = 4.9 kOe ($\mu_0 H_c$ = 0.49 T), $|BH|_{max}$ = 4.4 MGOe (35 kJ/m^3)) [117]. The rather low coercivities registered for these melt-spun alloys are caused by rough grain boundaries and discontinuous $CeFe_2$ grain boundary phase [118].

As reported by Y. L. Huang et al. [119], in the fabrication of anisotropic $Ce_2Fe_{14}B$ hot-deformed alloys, the addition of Pr-Cu low melting point eutectic alloy enhances deformability, the resulting magnets measuring remanent polarization J_r = 0.66 T, H_{ci} = 514 kA/m ($\mu_0 H_c$ = 0.64 T), $|BH|_{max}$ = 55 kJ/m^3. Sintered $Ce_2Fe_{14}B$ magnets fabricated from a mixture of $[Ce_{0.8}(Ho, Gd)_{0.2}]_{15.1}Fe_{75.7}B_{7.2}M_2$ (at.%; M = Co, Al, Cu, Nb) strip-cast flakes and 5 wt% Ce-rich alloy $Ce_{35.58}Fe_{53.47}Al_{10}B_{0.95}$ alloy with 826 °C melting point measured B_r = 7.567 kG (0.76 T), H_{cj} = 1.551 kOe ($\mu_0 H_c$ = 0.16 T) and $|BH|_{max}$ = 8.254 MGOe (65.68 kJ/m^3) [120].

The suppression of $CeFe_2$ through La substitution in $(La, Ce)_2Fe_{14}B$ melt-spun alloys improves magnetic properties, these alloys reaching double the energy product of $Ce_2Fe_{14}B$-based ones. M. Zhang et. al [95] used industrial La-Ce mischmetal to prepare $(La_{0.35}Ce_{0.65})_xFe_{14}B$ (x = 2.0, 2.4, 2.8, 3.2, 3.6 and 4.0) alloys, free of $CeFe_2$, with coercivity monotonically increasing with increasing x and energy product reaching a maximum of 8.29 MGOe (65.97 kJ/m^3) at x = 2.4 concentration. Q. M. Lu et al. [121] produced $(La_{0.3}Ce_{0.7})_3Fe_{14}B$ melt-spun ribbons with remanence of 6.86 kG (0.67 T), a coercivity of 5.51 kOe (μ_0H_c = 0.55 T), and $|BH|_{max}$ = 7.68 MGOe (61.12 kJ/m^3) and used SPS for hot compaction but the compacted magnets showed lower properties and required Nd-Cu low melting eutectic addition. Similar properties (H_{cj} = 345 kA/m (μ_0H_c = 0.43 T), J_r = 0.60 T, and $|BH|_{max}$ = 6.3 MGOe (50.13 kJ/m^3)) were obtained for $(Ce_{0.7}La_{0.3})_{2.5}Fe_{14}B$ melt-spun alloy [96].

Chemically non-uniform microstructures such as *multi-main-phase and core-shell grained magnet alloys* were shown to yield good magnetic properties. The multi-main-phase core-shell microstructures (with non-uniform rare-earth distribution within the grain and across multiple grains) are obtained by mixing and co-sintering RE-Fe-B powders with different rare-earth element constituent in their compositions (as in Ce-La or Y-Ce co-substituted sintered magnets [122]. This technique was adopted due to its cost benefit brought by the usage of fractions of non-critical rare-earth (such as Ce or/and La but also Y) RE-Fe-B powders and because better properties are obtained when alloy powders with different rare-earth content are co-sintered comparing to the substituted sintered alloy with homogenous distribution and the same average composition. [123, 87]. These heterogeneous microstructures are generated due to the rare-earth elements cross-diffusion during the liquid phase sintering. The method requires that powder fractions in the mix are optimised so that the resulting microstructure, although chemically non-uniform, shows uniform magnetization reversal [124, 125]. L. Zhang et al. [124] prepared dual main phase sintered magnets with $|BH|_{max}$ = 36.7 MGOe (292.05 kJ/m^3) for a composition with 45 wt. % Ce substituted in (Nd, Pr)-Fe-B alloy. J. Jin et al. [123] reported high energy product $|BH|_{max}$ = 42.2 MGOe (335.81 kJ/m^3) in sintered magnets with 36 wt. % La-Ce content, down from 48.9 MGOe (389.13 kJ/m^3) in the La-Ce-free (Nd, Pr)-Fe-B reference magnet.

Powder alloy mixtures containing mischmetal-based $R_2Fe_{14}B$ powder fractions have also been reported to yield energy products exceeding 34 MGOe (270.56 kJ/m^3) at mischmetal to total rare-earth content below 21.5 % [126].

Core-shell grained microstructures prepared by Y-Ce co-substitution in $Nd_2Fe_{14}B$ (Y-enriched core and (Nd, Ce)-enriched shell) where shown to produce high coercivities and enhance thermal stability [127]. Improved thermal stability and coercivity H_{cj} = 11.8 kOe (μ_0H_c = 1.18 T), remanence B_r = 13.7 kG (1.37 T) and energy product $|BH|_{max}$ = 45.6 MGOe (362.87 kJ/m^3) were obtained for Y-Ce co-substituted (9 wt%) $(Pr, Nd)_{30}Fe_{bal}M_{1.3}B_1$ magnets with core-shell grain structure obtained by co-sintering Y-containing and Y-free powders [128]. Core-shell grained microstructures are also obtained by grain boundary infiltration with low melting point alloys in both sintered and hot-deformed magnets. Grain boundary infiltration with Nd-Cu eutectic performed on hot-deformed $Ce_2Fe_{14}B$ magnets was shown to produce core-shell grains with Nd-rich shell and improve coercivity [129]. A similar effect was shown for $Ce_2Fe_{14}B$ sintered magnets reaching 30 kJ/m^3 at 20 wt % Nd-Cu addition [130].

Cerium valence manipulation

In the early X-ray and neutron diffraction work by Herbst and Buschow (referred to in Section 2.2) it was established that Ce is essentially tetravalent, lacking magnetic moment in the $Ce_2Fe_{14}B$ compound. Electronic structure investigations through X-ray absorption measurements indicated however that Ce is found in a strongly mixed-valent state which lacks a stable magnetic moment [131]. Also, no evidence of Ce^{3+} valence (which has a net magnetic moment) was found in $La_{2-x}Ce_xFe_{14}B$ (x = 0, 0.2, 0.4, ... 2) compounds [132]. Further, it was observed that Ce shifts from Ce^{4+} to the Ce^{3+} state in $Ce_2Fe_{14}BH_x$ hydride compounds [131], the shift to lower valence being correlated with the increased steric volume of the Ce atomic site which is due to the lattice expansion caused by interstitially absorbed hydrogen. Partially substituted $(Nd_{0.75}Ce_{0.25})_2Fe_{14}B$ features ordering tendencies on the two non-equivalent rare-earth crystallographic sites (*4f* and *4g*) [133]. Theoretical calculations predicted that in $(Nd_{1-x}Ce_x)_2Fe_{14}B$ Ce would preferentially occupy the *4g* site (larger in volume than *4f*) while in $(La_{1-x}Ce_x)_2Fe_{14}B$ Ce would go to the *4f* site [134]. Also by theory, it was predicted that in the $(Nd_{0.75}La_{0.25})_2Fe_{14}B$ compound La would prefer the *4g* site and produce a slight unit cell expansion [99]. According to E. T. Teatum et al. (1968) [135], the atomic radii of Nd^{3+}, Ce^{3+}, Ce^{4+} and La^{3+} atoms have the lengths of 1.821 Å, 1.825 Å, 1.715 Å and 1.877 Å respectively. In the $Nd_2Fe_{14}B$ cell, Ce^{3+} and Ce^{4+} would therefore substitute Nd on the *4g* and *4f* site respectively but site occupancy is still a matter of debate (reported data differs, be it from experiment - powder diffraction [136], single crystal [109] or theoretical modelling [134]. These findings motivated investigation and experimental work that tested if the transition toward the Ce^{3+} state could be achieved by making use of alloying additives to modify the coordination space environment. Lanthanum was shown to help shift the Ce valence in the $Nd_2Fe_{14}B$ structure through site steric volume increase due to its large atomic radius. For instance, by XPS (x-ray photoelectron spectroscopy) analysis of $[(Pr, Nd)_{1-x}(La, Ce)_x]_{2.14}Fe_{14}B$ ($0 \leq x \leq 0.5$) strip-cast alloys it was shown that Ce-La co-substitution shifts the valence of Ce to the Ce^{3+} state [137]. Others claim that Ga substitution for Fe also supports the Ce^{3+} state in the Φ-phase structure as indicated for the $Ce_{17}Fe_{78-x}B_6Ga_x$ (x = 0 - 1.0) melt-spun alloys [138].

Processing of Ce/La magnets by hydrogen treatment

Currently there is little data reported regarding the behavior of Ce and/or La containing $R_2Fe_{14}B$ magnet alloys when exposed to hydrogen treatment. A recent study [139] on the decrepitation behavior of $(La, Ce)_{31.5}Fe_{bal}B_{1.0}$ wt. % strip-cast alloy indicated that hydrogen absorption in this alloy was more sluggish, with a less exothermic effect, required higher gas pressure to decrepitate and overall absorbed less hydrogen comparing to its Nd-based counterpart.

This behavior was ascribed to the fact that, comparing to the Nd-based one, the (Ce, La)-based cast alloy showed higher oxygen content implying that the rare-earth-rich intergranular phase is more strongly oxidized which impedes hydrogen absorption.

However, an analysis of the cast alloy phase structure and microstructural features (phase configuration, intergranular material phase and elemental composition and distribution in the alloy etc.) is not provided so a complete picture of the correlation between the cast alloy characteristics and its response to hydrogen exposure cannot be drawn from the reported data.

Affinity to oxygen was shown to decrease with decreasing atomic radius for lanthanides [140], therefore Ce and La present at the grain boundaries make the alloy more sensitive to oxidation. For instance, the oxygen content was found to increase with x in $(Nd_{1-x}La_x)_{15.5}Fe_{77}B_{7.5}$ sintered magnets, reaching ~ 1.2 wt % at $x = 0.8$, twice comparing to the La-free sample [108].

Prior to the present work, only one research report was available in the literature regarding the HDDR treatment of Ce-based $R_2Fe_{14}B$ magnets [141]. The studied composition was $Ce_{13}Fe_{80}B_7$ and it was shown that anisotropic powders can be obtained through HDDR treatment but the as-treated powders showed very low coercivity (400 Oe ($\mu_0H_c = 0.04$ T)). The coercivity was ameliorated (up to 1450 Oe ($\mu_0H_c = 0.15$ T)) after the HDDR powder was reprocessed by grain boundary infiltration with $Ce_{72}Cu_{28}$, a low melting point eutectic alloy. The analysis of Ce and/or La containing $RE_2Fe_{14}B$ magnet alloys response to hydrogen treatment (HD and HDDR) is also a very effective tool in the material characterization routine. As emphasized also in section 2.3.3, it provides essential information regarding the composition and microstructure dependent hydrogen absorption/desorption behaviour and related evolutions (grain boundary processes - decompositions, diffusion, melting) which has implications for practical magnet production (for powder sintering but also for HDDR) and allows optimal alloy design and processing (composition choice, heat treatment regimes) to maximize magnetic properties.

3 Experimental techniques

This chapter details on the experimental work that was performed and lead to the results included in the book. It is divided in two sections, the first one referring to the synthesis of the alloys that were considered for study and the magnet fabrication while the second section contains the description of the investigation techniques that were used for the characterization of the obtained materials and magnets. The experimental work was carried out mainly at Fraunhofer IWKS in Hanau and Alzenau and TU Darmstadt while a part of the synthesis activity took place in an industry internship setting, at Magneti Ljubljana, in Slovenia. The EPMA measurements were performed in the Materials Engineering Department at KU Leuven, in Belgium.

3.1 Alloy synthesis and processing routes

Two processing routes have been adopted for the fabrication of permanent magnets using rapidly solidified $Nd_2Fe_{14}B$-based alloys:

(I) *Melt-spinning* followed by *hot-pressing* and *hot-deformation* of $(Nd_{1-x}R_x)_{13.6}Fe_{73.6}Co_{6.6}Ga_{0.6}B_{5.6}$ alloy with R = Ce, La and x = 0, 0.1, 0.2 ... 1 (for R = Ce) and x = 0, 0.1, 0.2 ... 0.5 (for R = La);

Alloy ingot samples of the above-mentioned compositions were prepared by arc-melting pure element ingredients and ferroboron (FeB) pre-alloy under protective argon atmosphere in an Edmund-Bühler AM 200 arc-melter. During arc-melting the melt is solidified on a water-cooled copper hearth. The obtained ingot samples (of 10 g) were flipped and re-melted several times for homogenization.

Melt-spun ribbons were produced in an Edmund-Bühler PA 500 melt-spinner (Figure 3.1) by induction melting the ingots in the evacuated and Ar back-filled chamber using boron nitride and quartz crucibles with round nozzle and ejecting the melt on the chilled copper wheel rotating at 30 m/s tangential speed.

Figure 3.1: Inside view of the melt-spinner chamber (a), quartz crucible during melting (b) and resulting melt-spun flakes (c).

A hydraulic press (Weber, 400 kN maximum load) was used for the uniaxial *hot-pressing* and *hot-deformation* processes (illustrated in Figure 3.2). Refractory steel dies with diameters of 13.5 mm and 22.5 mm were used for hot-pressing and hot-deformation respectively.

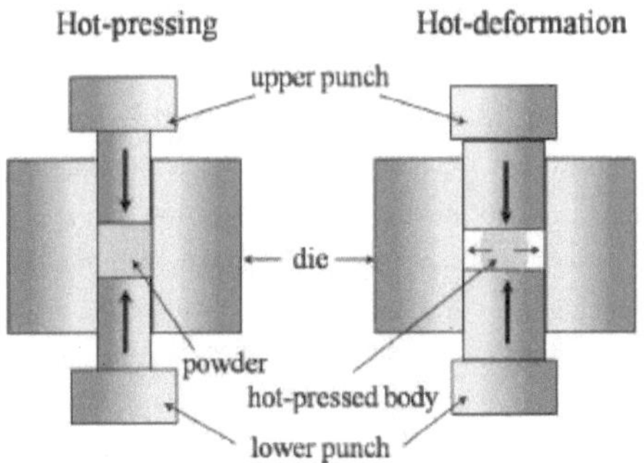

Figure 3.2: Simplified schematic representation of the assemblies used for hot-pressing and hot-deformation.

In order to prevent the sample material from getting stuck into the die, all inner walls of the die assembly are coated with a thin graphite layer lubricant. Fully dense isotropic magnets were produced by hot-pressing samples of 9 g of melt-spun powder. The process was conducted in vacuum (10^{-2} - 10^{-3} mbar) at 725 °C at a maximum force of 13 kN applied for 2 min. The hot-compacted body was subject to further processing by hot-deformation in protective Ar atmosphere at 750 °C, under a maximum compression force of 40 kN at a degree of deformation $\varphi - 1$ and a deformation rate $\dot{\varphi} = 0.01\,\mathrm{s}^{-1}$. The degree of deformation φ is calculated according to the formula:

$$\varphi = \ln\left(\frac{h_0}{h_{final}}\right) = \ln\left(\frac{d_0^2}{d_{final}^2}\right) \tag{3.1}$$

The final height of the deformed sample (h_{final}) is given by the following expression:

$$h_{final} = \frac{h_0 \times d_0^2}{d_{final}} \tag{3.2}$$

where h_0 and d_0 are the initial height and the diameter of the of the hot-compacted body and d_{final} is the diameter of the die used for hot-deformation.

(II) *Strip-casting* followed by room temperature *hydrogen decrepitation* (HD) and *dynamic hydrogenation disproportionation desorption recombination* (d-HDDR) of $(Nd_{1-x}Ce_x)_{15}Fe_{79}B_6$ base composition where x = 0, 0.1, 0.2, ... 0.6;

The strip-cast flakes were obtained by induction melting pure elements and FeB pre-alloy (10 kg batch) in Ar atmosphere and pouring the melt through a round nozzle (1 cm in diameter) on a rotating chilled copper disk at 7 m/s tangential speed. The process was conducted at Magneti Ljubljana (Slovenia) using an in-house built strip-caster.

The strip-cast flakes were subjected to HD treatment, the process being performed in a stainless steel autoclave at room temperature in 1 bar H_2 pressure. Batches of ~ 50 g of strip-cast sample material were placed into the autoclave which was sealed and connected to a vacuum and gas feed (Ar and H_2) pipe system. In order to protect the material from oxidation, the autoclave was evacuated and flushed with Ar three times prior to allowing the H_2 reaction gas into the vessel. Hydrogen was introduced into the autoclave up to a pressure of 1 bar which was kept constant by filling to compensate for the absorbed gas until the manometer mounted on the autoclave showed no change in hydrogen pressure reading (indication that the material no longer absorbs gas) then the hydrogen valve was closed and the vessel was placed in safe storage for 24 hours to ensure that the decrepitation reaction is complete. After 24 hours hydrogen was evacuated and the autoclave was transferred to the Ar glove-box for the decrepitated material to be crushed and stored under inert atmosphere to prevent oxidation.

The HDDR treatment applied on the decrepitated powders was conducted in constant H_2 pressure of 30kPa with a temperature plateau at 780 °C and 840 °C for disproportionation and 840 °C for desorption - recombination. A schematic representation of the d-HDDR treatment profile featuring processing parameters (H_2 pressure, temperatures and dwell times) together with the specific microstructures which form at the different stages of reaction are presented in Figure 3.3. The HDDR system is automated and the process parameters (hydrogen pressure, temperature, hydrogen desorption rate) are computer-controlled.

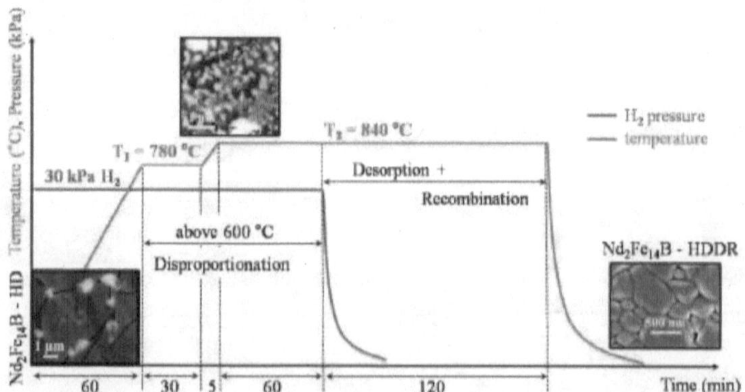

Figure 3.3: Treatment profile of the employed *d*-HDDR process and SEM details showing the starting decrepitated material and illustrating the phase and morphology transformations of the Φ-phase grain (starting from the decrepitated microstructure, lower left image) at the disproportionation and recombination stages.

Throughout the entire process the samples are handled under protective atmosphere. The sample loading is done in the Ar glove-box where the decrepitated powder is filled into the sample holder which is placed and sealed into a tubular treatment chamber fabricated of a highly corrosion-resistant refractory alloy. The sealed assembly is transferred from the Ar glove-box to the HDDR system and inserted into the furnace where it is surrounded by a quartz tube which is evacuated during the treatment. For disproportionation, at constant H_2 pressure, the temperature is ramped then maintained in two annealing steps as shown in Figure 3.3. The hydrogen pressure is decreased to initiate the recombination, then the system is pumped for the vacuum annealing process to complete the recombination reaction. The quartz tube is flushed with nitrogen gas to speed up the cooling at the end of the process when the furnace is removed. For the HDDR powder to be removed and stored the sample chamber tube assembly is disconnected from the HDDR system and transferred to the Ar glove box.

Strip-cast alloys with the composition $(Nd_{1-x}La_x)_{12.5}Fe_{81.2}B_{6.3}$ at. % (x = 0, 0.1, ... 0.3) were also produced and selected data is presented in this work for comparative analysis with the other produced alloys particularly with regards to the as-cast phase structure, to the hydrogen absorption-desorption characteristics and to the Curie transition temperatures of the Φ-phase (figures are introduced in the Annex). In order to test the effect of La substitution, the choice of a reference composition with a lower rare-earth content is a more suitable one since in contrast to Ce, La is expected to not form secondary phases in the alloy and mainly segregate at the grain boundaries due to its limited substitutional solubility in the Φ-phase. For the Ce-containing alloys, rare-earth-rich overstoichiometric compositions were shown to produce better magnetic properties. They promote the formation of higher fractions of $CeFe_2$ phase in alloys with high Ce concentrations where a rare-earth-rich phase does not form. If uniformly distributed at the grain boundaries, $CeFe_2$ can act as an intergranular spacer for the magnetic decoupling of the Φ-phase grains and enhance coercivity similarly to the rare-earth-rich phase.

3.2   Sample characterization techniques

3.2.1   Quantitative elemental analysis

Inductive coupled plasma optical emission spectroscopy (ICP-OES)

The elemental composition of the strip-cast alloys was determined by ICP-OES measurements performed on a PerkinElmer - Optima 8300 instrument. For analysis, samples were crushed into powders and diluted in a mixture of hydrochloric acid (HCl) and nitric acid (HNO_3).

ICP-OES is an elemental analysis method based on optical spectroscopy [142] that allows the detection and measurement of ~ 70 chemical elements. The samples which are liquids or solids dissolved in liquids are vaporized and atomized in plasma.

The chemical bonds are broken through atomization and ionization and the resulting free atoms and ions of the species present in the sample are excited and emit light. Each chemical element emits a characteristic spectrum in the ultraviolet and visible region and the light intensity at one of the characteristic wavelengths is proportional to the concentration of that element in the analyzed sample.

Inert gas fusion technique

The oxygen content was determined by the inert gas fusion technique using a LECO ONH836 Oxygen/Nitrogen/Hydrogen elemental analyzer.

The inert gas fusion (IGF), also commonly termed as melt extraction (ME), is a destructive analysis technique which allows the measurement of gas forming elements (C, H, O, N and S) present in solid materials in quantities ranging from ppm (parts per milion) to percentage levels. It involves the melting (fusion) of the sample in a inert gas atmosphere furnace at high temperatures in a single-use graphite crucible. After purging with an inert gas (He or Ar), a high current is passed through the crucible generating temperatures above 2500 °C. The process breaks the physical and chemical bonding between the gasses and metals to dissociate the gasses and sweep them from the fusion area with the inert gas carrier. Any gases generated in the furnace (CO, CO_2, N_2 and H_2) are released into the flowing gas stream which is directed to the infrared detectors (for O measured as CO and CO_2) or thermal conductivity for N and H. The extraction, separation, detection and measurement of oxygen, nitrogen and hydrogen are done simultaneously.

3.2.2 Phase structure and crystal properties analysis

X-Ray Diffraction (XRD)

For all studied alloys, at different processing stages, the structural state and phase composition were determined by X-ray powder diffraction (XRD) measurements performed on a Panalytical Empyrean instrument with Co-K_α radiation. Pattern profile fitting was done using the HighScorePlus software.

XRD [142, **143**] is an essential technique for the structural analysis of materials. It allows phase identification by giving information about the specific inner crystal geometry of bulk materials, powders and thin films. The structural state (orientation, degree of crystallinity, crystallite size, lattice parameters, internal stress and strain) can also be determined from XRD data.

The structure of crystalline materials is composed of a periodic array of atoms being characterized by the long-range order of the atomic arrangement. The basic structural unit that contains the maximum symmetry specific to a crystal structure is the unit cell which by repetition defines the 3D crystal lattice.

Depending on shape and symmetry, there are seven types of unit cells (triclinic, monoclinic, orthorhombic, tetragonal, cubic, trigonal and hexagonal) determined by the axial relations that describe them. As in the case of monoclinic, orthorhombic, tetragonal and cubic unit cells different atomic arrangements and packing are possible, 14 lattice types can be distinguished. Crystal lattices consist of parallel atomic planes that are interspaced by the constant distance d. The atomic planes intersect the unit cell and can be identified using the Miller indices (h k l) which are the reciprocals of the fractional intercepts with the crystallographic axes of the 3D lattice. Diffraction occurs when a radiation beam of wavelength comparable or smaller than the lattice constant is scattered by the atoms in the lattice, producing constructive interference at specific angles. Constructive interference requires that the scattered beams come out in phase, which means that the path difference between the incident and outgoing rays equal an integer number of wave lengths. For a family of atomic (h k l) planes with a distance d_{hkl} between them, the condition for constructive interference to occur is the fulfillment of Bragg's law:

$$2d_{hkl}\sin\theta = n\lambda \tag{3.3}$$

where d_{hkl} is the interplanar spacing, θ is the angle between the incident X-ray beam and the atomic planes, λ is the wavelength of the incident beam, and n is an integer number corresponding to the order of diffraction.

Bragg's law calculates the angle where the X-rays scattered by parallel planes of atoms will interfere constructively to produce a diffraction peak. Consequently, a family of crystalline planes produces a diffraction peak only at a specific angle 2θ. A schematic representation of Bragg reflections from the crystal lattice planes is shown in Figure 3.4.

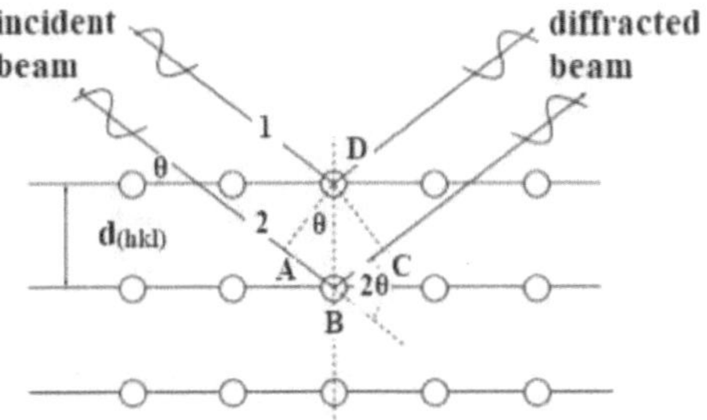

Figure 3.4: Bragg reflections from a set of lattice planes in a crystal, redrawn from ref. [144].

The unit cell parameters in the tetragonal crystal lattice, the example of interest for this work, are described by the a = b ≠ c relation, the angles between them being α = β = γ = 90°.

The lattice parameters of the tetragonal unit cell can be calculated using the expression that relates them to the interplanar distance d and the hkl Miller indices as follows:

$$\frac{1}{d_{hkl}^2} = \frac{h^2 + k^2}{a^2} + \frac{l^2}{c^2}$$

(3.4)

As shown above, the interplanar distance d is dependent on the diffraction peak position and is calculated using the Bragg's law, with the angle of diffraction θ expressed in radians.

3.2.3 Microstructure and related chemical composition analysis

Scanning Electron Microscopy (SEM)

In this work, SEM analysis and EDX mappings were performed on a Zeiss Merlin high resolution field emission gun (FEG) instrument equipped with a Gemini II column. For the analysis of the melt-spun ribbons surface as well as for the fracture surface of the hot-deformed magnets there was no particular preparation required. Sample preparation consisting of grinding and mechanical polishing was performed for the bulk materials while the powder samples were analyzed either in free form or after embedding, fine grinding and mechanical polishing followed by ion-milling (this procedure was applied also for the strip-cast flakes cross-section).

SEM [142] allows the caption of high magnification images that illustrate the surface topography of the analyzed material. It can also give information about the elemental composition near the surface of the specimen. The technique involves the scanning of the sample surface with an electron beam in vacuum which causes the emission of Auger electrons, secondary electrons (SE), backscattered electrons (BSE), photons and characteristic X-rays (Figure 3.5(a)).

The distinction between the secondary electrons and the backscattered electrons is made conventionally as a function of their energy. If the emitted electron has an energy lower than 50 eV, it is considered as being a secondary electron. The backscattered electrons have energies higher than 50 eV, comparable to the energy of the primary beam. The energy difference between the secondary electrons and the backscattered ones can be explained by the different mechanisms that produce them. The secondary electrons are emitted from the specimen when high energy electrons (from the primary beam or backscattered electrons) suffer an inelastic collision with the electrons of the atoms in the sample while the backscattered electrons are a result of elastic collisions.

Surface topography images are produced by the secondary electrons emitted from the sample. The contrast is obtained from the difference in height on the surface because the generation of secondary electrons depends on the angle of incidence between the beam and the sample (Figure 3.5(b)). The chemical composition contrast is realized by the backscattered electrons (Figure 3.5(c)). The measured signal depends on the atomic number Z of the elements which are contained in the sample material. The heavier the elements (larger Z) the higher becomes the probability of backscattered electrons emission so the registered signal will be stronger and corresponding sample region will appear brighter.

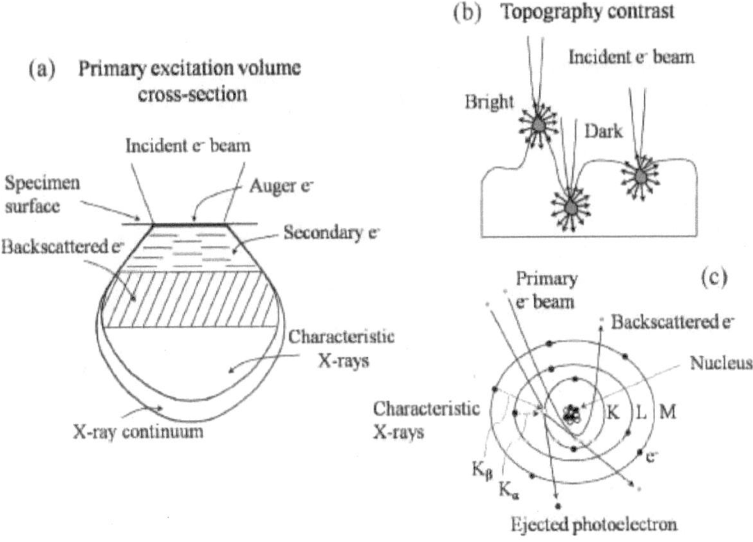

Figure 3.5: Incident electron beam - sample interaction volume (a), topography contrast (b) and backscattered electrons and characteristic X-rays generation (c).

Energy-Dispersive X-Ray Spectroscopy (EDX)

Characteristic X-rays are emitted when a primary electron removes an electron from the inner shells of an atom in the material generating the transition of the other electrons from the higher energy levels (Figure 3.5(c)). When the atom comes back to the fundamental state it emits a photon of a certain energy which is specific for each chemical element thus allowing its identification [142].

Electron Probe X-Ray Microanalysis (EPMA)

The chemical composition of the strip-cast flakes constituting phases was checked at the different treatment stages by wavelength-dispersive spectroscopy (WDX) measurements performed on an Jeol JXA-8530F electron probe micro-analysis instrument (EPMA).

Quantified elemental mappings of the strip-cast-flakes cross-sectional polished surfaces were thus obtained. Map quantifications for Nd, Ce, Fe, B, O, Al and Si were done relatively to NdF_3, CeF, pure Fe, BN, cuprite (Cu_2O), pure Al and pure Si standards.

The EPMA [142] device features the imaging capabilities of the SEM (SE and BSE detection) along with quantitative elemental analysis using both energy- and wavelength-dispersive X-Ray spectrometry, typically incorporating an EDX system together with several WDX ones.

EDX and WDX both measure the characteristic X-Rays emitted from the specimen but they rely on different detection mechanisms. While EDX uses the photoelectric absorption of X-Rays is a semiconductor crystal (Si or Ge), WDX is based on the diffraction of incident X-Rays on crystals with different d-spacing in order to detect a wide range of wavelengths. Compared to energy-dispersive spectrometers, wavelength-dispersive spectrometers come with far better X-Ray resolution and detection limits allowing the detection and quantitative measurement of elements as light as boron.

3.2.4 Thermal analysis

Differential scanning calorimetry (DSC)

The thermal behavior (crystallization, melting temperatures) was analyzed for melt-spun and strip-cast alloys using a Netzsch 404 F1 Pegasus High-Temperature DSC instrument. The powder samples were heated above the melting point at 10 K/min heating rate.

DSC is a thermo-analytical technique [**145**, **146**] allowing the detection of change in physical properties of a sample along with temperature against time. It involves the measurement of the energy transferred as heat to or from the sample material (absorbed heat marks an endothermic effect while the radiated heat signals an exothermic effect) during a physical or chemical change in constant pressure conditions. The thermal response of the sample material is compared to a reference which does not undergo any thermal events over the operating temperature range while the same temperature is maintained in the sample and reference throughout the analysis. The results are influenced by the particle size and the packing condition of the sample.

Differential Thermal Analysis in H_2 atmosphere (H-DTA)

The strip-cast alloys hydrogen absorption-desorption behavior as a function of temperature was analyzed through differential thermal analysis (DTA) measurements using a home built device.

The purpose of the measurements is to identify the effects of Ce substitution on the hydrogen absorption, Φ-phase matrix disproportionation, hydrogen desorption and Φ-phase recombination reactions, for a better understanding of the way the replacement of Nd with Ce changes the behavior of the alloys during the decrepitation and HDDR treatment.

The strip-cast flakes were crushed into powder under protective Ar atmosphere in the glovebox in order to prevent oxidation and samples of 1 g were prepared for each measurement. The hydrogen absorption-desorption measurements consisted of three successive steps:

(i) The powder surface reactivation - the powder was heated up to 300 °C under constant high vacuum (below 10^{-4} mbar), maintained for 5 minutes then let cool to room temperature;
(ii) The hydrogen absorption scan - heating to 900 °C in 1 bar H_2 at 10 K/min heating rate then cooling to room temperature;
This measurement is performed in order to observe the interstitial absorption of hydrogen and the high temperature disproportionation of the hydrogenated Φ-phase matrix.
(iii) The hydrogen desorption scan - heating to 900 °C at 10 K/min heating rate under constant vacuum pumping conditions.
This measurement step is performed in order to observe the thermal desorption of hydrogen and the high temperature recombination of the Φ-phase matrix.

3.2.5 Magnetic measurements

The initial magnetization curves and the hysteresis loop measurements of the melt-spun and HDDR powder samples were carries out at room temperature using a Quantum Design PPMS-VSM instrument (Physical Property Measurement System - Vibrating Sample Magnetometer). The sample preparation consisted in mixing ~ 50 mg of powder material in liquid paraffin wax and sealing the mixture in an aluminum capsule. The HDDR samples were aligned in a 2 T magnetic field prior to the hysteresis loop measurements which were performed both parallel (magnetic easy axis) and perpendicular (hard axis) to the orientation direction. During alignment the capsule is heated to melt the wax and allow the powder particles to move and orient on the direction of the applied field. Melt-spun powders are isotropic therefore alignment is not necessary in their case. The magnetic moment registered on the hysteresis loops was converted to magnetic polarization taking into account the theoretical density of the $Nd_2Fe_{14}B$ compound (7.5 g/cm^3) and a shape demagnetization factor of 1/3. The demagnetization curves for the hot-pressed and hot-deformed magnets produced for this work were measured on a PERMAGRAPH® Magnet-Physik Dr. Steingroever GmbH instrument. Prior to the demagnetization curve measurement the magnets were magnetized in a magnetization coil connected to an impulse magnetizer. The Curie transition temperatures of the melt-spun powder samples were determined from magnetization vs. temperature M(T) curves measured up to 1100 K with 4 K/min heating rate in a 0.01 T applied field on a 7400 Series Lake Shore Vibrating Sample Magnetometer (VSM).

On the VSM device the measurement is performed in an open magnetic circuit and the sample is magnetized by electromagnets or by a superconducting magnet in the case of the PPMS-VSM instrument (open loop setup can be seen in Figure 3.6 (a)). The VSM measurement involves the vertical oscillation of the sample at constant frequency and amplitude thus inducing in the pick-up coils an alternating electromotive force which is proportional to the magnetic moment of the sample (as implied by Faraday's law of induction which states that a time-varying magnetic field will produce an electric field).

For the magnetic measurement using the hysteresisgraph the sample is centered inside the measuring coil and with its planar parallel surfaces in close contact with the magnetic poles (closed loop, Figure 3.6 (b)). The demagnetization factor in this case is zero.

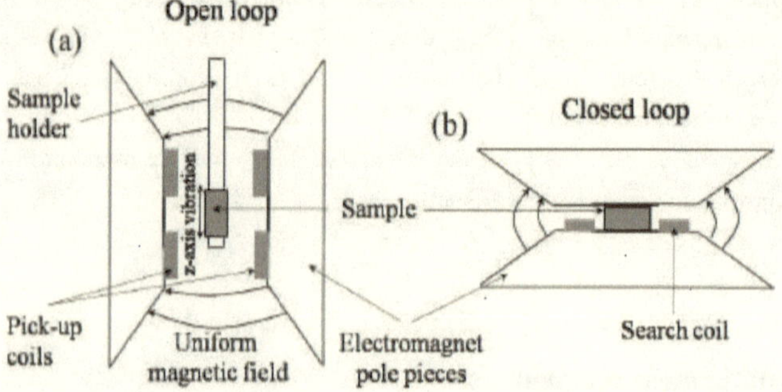

Figure 3.6: **Schematic of the magnetic measurement setup in open loop - vibrating sample magnetometry (a) and closed loop - hysteresisgraph (b).**

4 Results and discussion

In this chapter, the focus is on the influence of Ce and La substitution in $Nd_2Fe_{14}B$-based alloys obtained by rapid solidification techniques (melt-spinning and strip-casting) and processed by specific magnet fabrication methods. Following these two adopted processing routes described in the experimental section (Chapter 3), this chapter is divided in two sections. The results presented and analyzed in this work represent the basis of two research articles, published by the author in references [147, 148].

4.1 Ce and La substitution in $Nd_2Fe_{14}B$-based melt-spun alloys

This subchapter contains a comparative study on the influence of Ce, respectively La substitution on the phase composition, the crystal properties of the solidified Φ-phase, the microstructure and the magnetic properties of the $(Nd_{1-x}R_x)_{13.6}Fe_{bal}Co_{6.6}Ga_{0.6}B_{5.6}$ at. % (x = 0, 0.1, 0.2, ... 1 for R = Ce and x = 0, 0.1, 0.2, ... 0.5 for R = La) melt-spun alloys. The influence of substitution on the thermal behavior of the melt-spun alloys and their response to hot-pressing and hot-deformation is also discussed.

The alloy taken for reference, composition $Nd_{13.6}Fe_{73.6}Co_{6.6}Ga_{0.6}B_{5.6}$ at. %, is of particular technological interest being commercially available as MQU-F grade, produced by Magnequench. The magnetically isotropic melt-spun material can be used for polymer bonded magnets fabrication [149]. The MQU-F grade composition is also optimized for the production of permanent magnets by hot-pressing and hot-deformation. The hot-pressed magnet also is isotropic but through plastic deformation of the grains (elongation normal to the press direction) magnetic anisotropy is obtained as c-axis crystallographic alignment (texture) is induced [150, 151, 152].

SEM micrographs of the MQU-F melt-spun powder and the room-temperature hysteresis loop measured on this powder can be seen in Figure 4.1. The images of the ribbon fragments reveal smooth surfaces and an uniform grain structure. The powder measured a coercivity μ_0H_C of 1.90 T and a remanence B_r of 0.80 T.

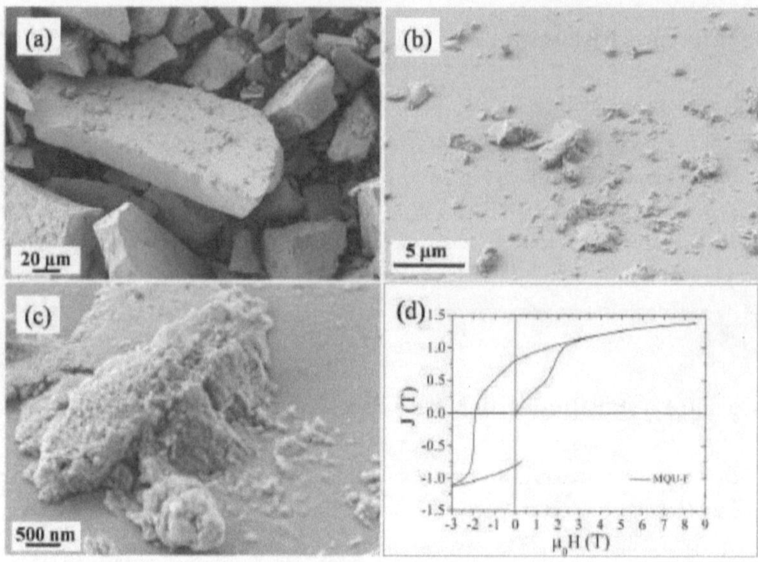

Figure 4.1: SEM images of the commercially available MQU-F grade melt-spun powder showing ribbon fragments (a) and magnified details on the ribbon surface (b, c) and the room-temperature hysteresis curve measured on the powder (d).

In a first synthesis step, alloy ingots of composition $(Nd_{1-x}R_x)_{13.6}Fe_{bal}Co_{6.6}Ga_{0.6}B_{5.6}$ at. % (x = 0, 0.1, 0.2, ... 1 for R = Ce) and (x = 0, 0.1, 0.2, ... 0.5 for R = La) were prepared by arc-melting. The BSE micrographs of the $(Nd_{1-x}Ce_x)_{13.6}Fe_{74.2}Co_{6.6}B_{5.6}$ ingots polished surfaces are shown in Figures 4.2, 4.3 and 4.4 (for x = 0, 0.5 and 1 respectively) together with EDX elemental mappings and point analysis spectra. They reveal the formation of the hard magnetic Φ-phase for all compositions (dark gray regions with corresponding EDX spectra in spot 3 as indicated in all the three figures). The absence of boron is caused by the lower detection efficiency of its characteristic X-rays by EDX analysis. As a result of low solidification rate attained through arc-melting synthesis, the dendritic segregation of primary α-Fe phase can be observed in each sample (black contrast). This type of microsegregation is also called coring and occurs in non-equilibrium cooling conditions. The Nd-rich phase is visible in bright contrast in Figure 4.2 (Spot 2). At x = 0.5 Ce substitution ratio (Figure 4.3), an additional phase segregates in the alloy, the Laves-type $CeFe_2$ phase which becomes predominant at the intergranular regions (visible in light gray contrast) while the Nd-rich phase can be still observed in brighter contrast at the Φ-phase grains edges. The $CeFe_2$ phase which is known to form in the Ce-Fe-B alloy system according to the phase diagram [89] has a CuMg-type face-centered cubic (FCC) crystal structure and is paramagnetic at room temperature showing a Curie transition at 230 K [153]. Phase segregation intensifies at complete replacement of Nd with Ce in the alloy composition where one may observe that the Φ-phase proportion is significantly diminished with the increase in proportion of the $CeFe_2$ phase (Figure 4.4).

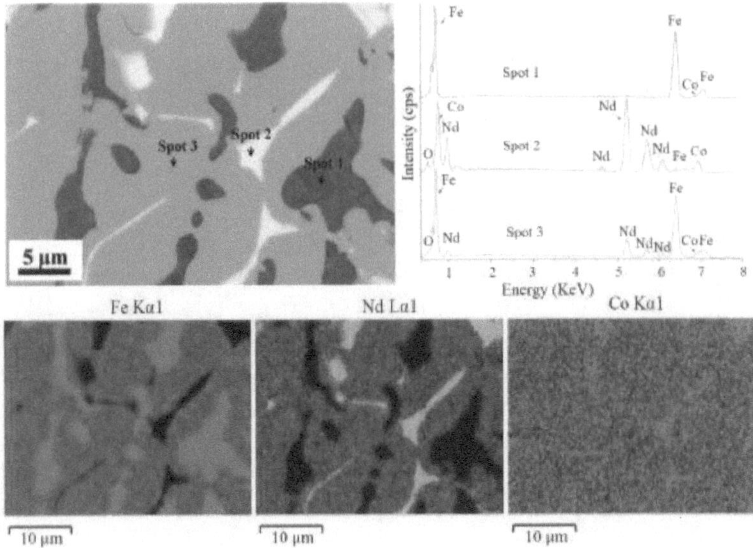

Figure 4.2: BSE image, EDX elemental mappings and spot analysis spectra of $Nd_{13.6}Fe_{74.2}Co_{6.6}B_{5.6}$ (at. %) ingot prepared by arc-melting.

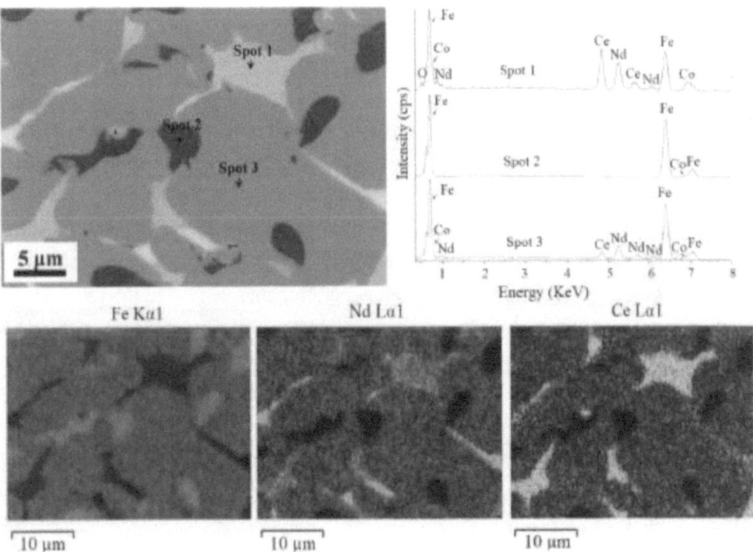

Figure 4.3: BSE image, EDX mappings and spot analysis spectra of the $Nd_{6.8}Ce_{6.8}Fe_{74.2}Co_{6.6}B_{5.6}$ (at. %) sample ingot prepared by arc-melting.

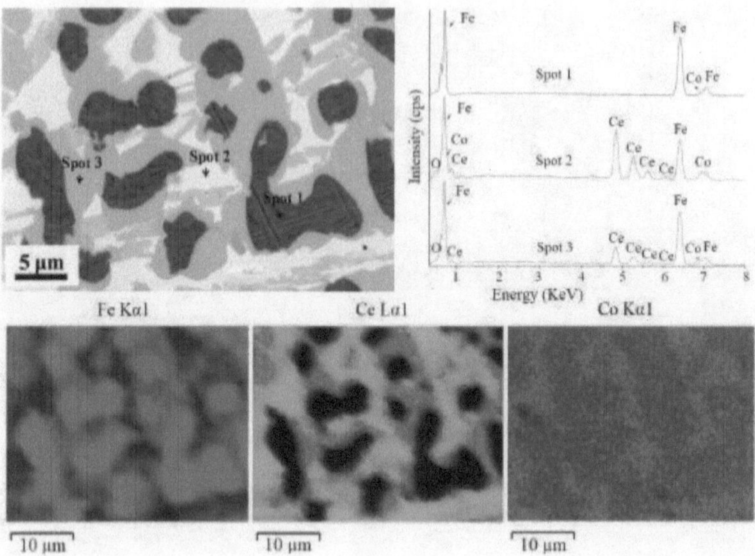

Figure 4.4: BSE image, EDX mappings and spot analysis spectra of the $Ce_{13.6}Fe_{74.2}Co_{6.6}B_{5.6}$ (at. %) sample ingot prepared by arc-melting.

4.1.1 Phase composition and crystal properties

The alloy ingots of composition $(Nd_{1-x}R_x)_{13.6}Fe_{bal}Co_{6.6}Ga_{0.6}B_{5.6}$ at. % (x = 0, 0.1, 0.2, ... 1 for R = Ce) and (x = 0, 0.1, 0.2, ... 0.5 for R = La) that were produced by arc-melting were used for the preparation of melt-spun ribbons without a prior homogenization heat treatment.

The ingots were molten by induction and the alloy ribbons were spun at relatively high wheel speed (30 m/s) which ensures a high cooling rate for direct quenching into nanocrystalline state. The oxygen content of the melt-spun alloys was determined for the reference and the Ce containing melt-spun alloys by means of inert gas fusion measurements. It was found to range from 200 to 600 ppm and did not correlate with the Ce concentration.

It is important to note here that the Ce and La atomic fractions refer to their content in the alloy nominal composition. The Ce and La atomic fractions within the hard magnetic Φ-phase are not discernable for the alloys because these elements also participate in the formation of the intergranular phases due to migration to the grain boundaries and phase segregations occurring during solidification. Powder samples were measured by XRD for all the prepared melt-spun ribbon alloys. Patterns corresponding to several compositions from $(Nd_{1-x}Ce_x)_{13.6}Fe_{73.6}Co_{6.6}Ga_{0.6}B_{5.6}$ sample series can be seen presented in Figure 4.5.

Measurements show that the tetragonal hard magnetic Φ-phase crystallized over the entire Ce concentration range while a single-phase composition (the intergranular phase is below the detection limit of XRD) was obtained up to x = 0.7 Ce fraction where weak reflections characteristic to the cubic $CeFe_2$ Laves-type phase appear on the patterns. At higher Ce concentrations the patterns indicate stronger reflections therefore increasing amounts of $CeFe_2$ segregated phase.

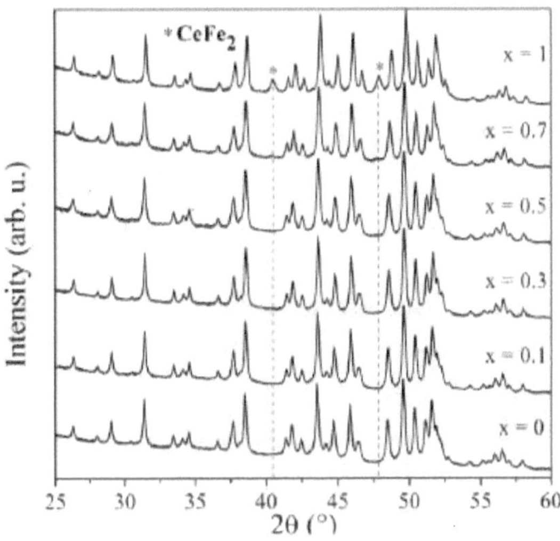

Figure 4.5: X-ray diffraction patterns of the $(Nd_{1-x}Ce_x)_{13.6}Fe_{73.6}Co_{6.6}Ga_{0.6}B_{5.6}$ (at. %) (x = 0, 0.1, 0.3, 0.5, 0.7 and 1) melt-spun alloys in as-quenched processing state.

The Φ-phase lattice constants (unit cell volume and parameters c and a) were derived by XRD pattern profile fitting. The variation of the lattice constants and tetragonality factor c/a within the alloy Ce concentration range is presented in Figure 4.6. It can be seen that the lattice parameters, the c/a tetragonality factor and the unite cell volume show a decreasing linear trend with increasing Ce concentration. Lattice contraction indicates the gradual increase of Ce concentration in the Φ-phase.

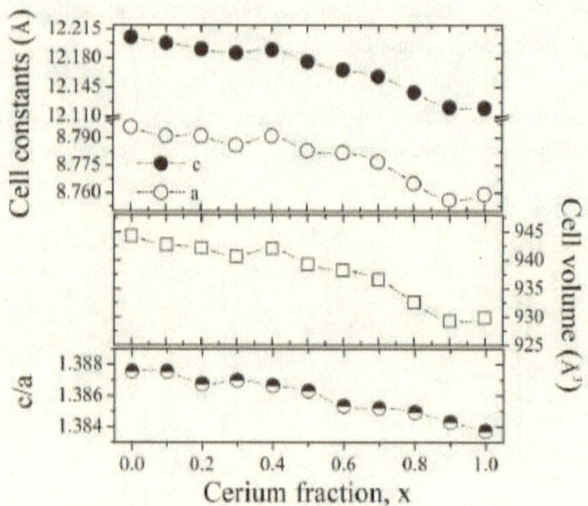

Figure 4.6: Unit cell parameters c and a, volume and tetragonality factor (c/a) of the Φ-phase plotted against Ce nominal content for $(Nd_{1-x}Ce_x)_{13.6}Fe_{73.6}Co_{6.6}Ga_{0.6}B_{5.6}$ as-spun alloys.

The XRD patterns measured on the La substituted as-quenched ribbons are introduced in Figure 4.7. All samples show crystallinity and the patterns indicate that the alloys are composed of the tetragonal Φ-phase and appear to be single-phase up to x = 0.4 La substitution ratio. At higher La concentration the peritectic segregation of α-Fe and Nd_2Fe_{17} cubic phases occurred as it can be seen from the additional peaks marked on the diffraction pattern of the sample with x = 0.5 La atoms ratio. The segregation of α-Fe and Nd_2Fe_{17} in the alloy indicates that La alters the solidification path (the peritectic segregation occurs already at high temperature upon cooling) because its solubility in the $Nd_2Fe_{14}B$ phase is limited and it does not form compounds with Fe (there is no compound formation in the Fe-La phase system [154]). Additionally, with increasing La fractions relative to Nd, a progressive broadening of the Φ-phase diffraction peaks can be observed on the patterns. The peak broadening suggests a decrease in crystallite size so one may further interpret that with La substitution the growth of the Φ-phase is restricted, nucleation being favored instead.

The XRD pattern of a melt-spun sample of composition $La_{13.6}Fe_{73.6}Co_{6.6}Ga_{0.6}B_{5.6}$ produced in the same conditions as the other alloys can be seen in Figure 4.8. This sample consists mainly of α-Fe as the peaks labeled on the pattern suggest. The other major peaks marked on the pattern correspond to a La-rich phase. There is no strong indication of the formation of the Φ-phase in this sample.

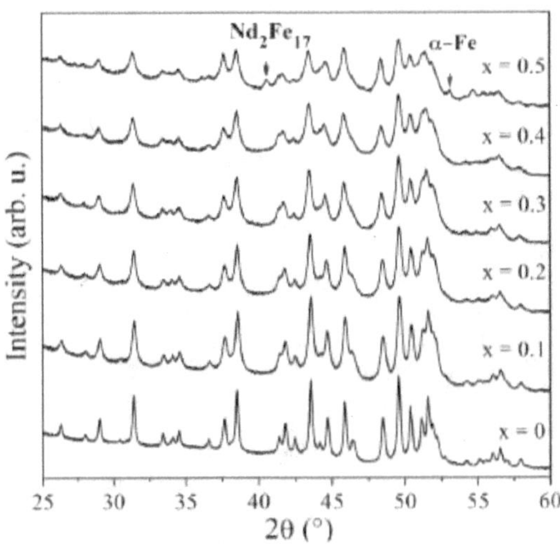

Figure 4.7: X-ray diffraction patterns of the $(Nd_{1-x}La_x)_{13.6}Fe_{73.6}Co_{6.6}Ga_{0.6}B_{5.6}$ (x = 0, 0.1, 0.2, ... 0.5) melt-spun alloys in as-quenched processing state.

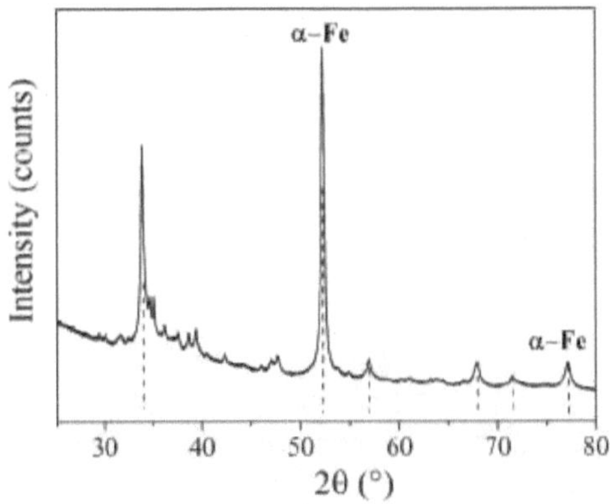

Figure 4.8: X-ray diffraction pattern of the $La_{13.6}Fe_{73.6}Co_{6.6}Ga_{0.6}B_{5.6}$ melt-spun alloy.

Figure 4.9 introduces the plot of the lattice constants and the unit cell tetragonality factor of the Φ-phase, calculated for the prepared $(Nd_{1-x}La_x)_{13.6}Fe_{73.6}Co_{6.6}Ga_{0.6}B_{5.6}$ as-spun ribbons. The cell parameters trend observed in this case shows that with increasing fractions of substitution for Nd, La progressively expands the unit cell of the Φ-phase.

The unit cell expansion occurs preferentially along the *c*-axis, being associated with an increasing tetragonality factor (*c/a*) (or an increasing aspect ratio) as the *a* parameter is varying in a narrow value range. The unit cell volume shows in their case a non-monotonic increasing trend with increasing La concentrations, a volume decrease being registered for the melt-spun sample with x = 0.1 La concentration. The unit cell volume reduction at x = 0.1 La fraction may suggest that, at low concentrations, La is expelled from the Φ-phase to the intergranular phase.

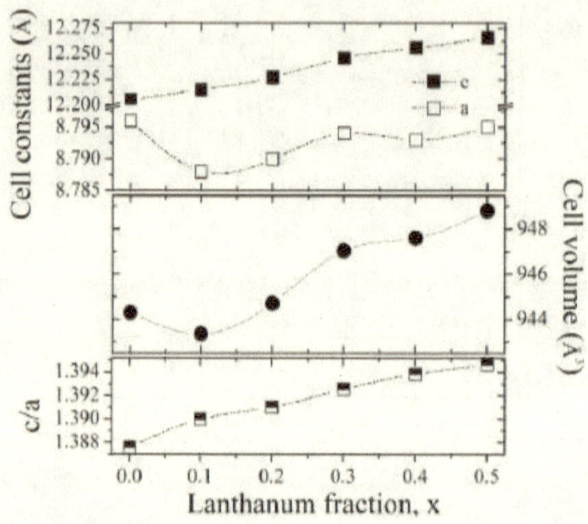

Figure 4.9: Unit cell parameters *c* and *a*, volume and tetragonality factor (*c/a*) of the Φ-phase plotted against La nominal content for $(Nd_{1-x}La_x)_{13.6}Fe_{73.6}Co_{6.6}Ga_{0.6}B_{5.6}$ as-spun alloys.

The La atom has a significantly larger radius comparing to Nd and it is partially soluble in the Φ-phase at high temperatures but it can be retained within its structure when high cooling rates as attained during melt-spinning are applied. For Ce on the other hand, the atomic radius varies between its two valence states and, comparing to Nd, Ce^{4+} is significantly smaller while Ce^{3+} is larger [135]. Therefore, if in a mixed-valence state, Ce^{3+} would be expected to fit in the larger 4*g* sites while Ce^{4+} will tend to the 4*f* sites of the $Nd_2Fe_{14}B$ unit cell. However, for powder and monocrystalline samples, the reported data does not converge (compared to alloy synthesis, single crystal synthesis involves very slow diffusion rates). Powder neutron diffraction measurements indicated that Ce preferentially occupies the smaller Nd 4*f* sites and forms a continuous substitutional solid solution in $Nd_2Fe_{14}B$ phase [136] while single crystal data from theoretical calculations and experiment showed a slight preference of Ce for the Nd 4*g* sites [134, 109].

On the other hand, La is partially soluble in the $Nd_2Fe_{14}B$ phase and is predicted by theoretical calculations to prefer the larger Nd 4g sites. The XRD data and related observations in this work are consistent with previously reported theoretical calculations which predicted the lattice expansion and also with the observed tendency of La atoms to segregate at the grain boundaries rather than enter the Φ-phase structure unless rapid quenching is employed [99].

The results show that high solidification rates are necessary to retain larger fractions of La in the Φ-phase structure and suppress the α-Fe segregation. At the slower cooling rates attained when strip-casting is used for alloy synthesis, α-Fe segregations occur in the cast structure at low La concentrations as indicated by XRD analysis of $(Nd_{1-x}La_x)_{12.5}Fe_{81.2}B_{6.3}$ at. % (x = 0, 0.1, ... 0.3) strip-cast alloys (patterns in Figure 6.1, in the Annex section).

4.1.2 Microstructure analysis

Figure 4.10 contains SEM images showing the microstructure on the surface of the $(Nd_{1-x}R_x)_{13.6}Fe_{73.6}Co_{6.6}Ga_{0.6}B_{5.6}$ melt-spun ribbons with x = 0, x = 0.1 Ce and 0.1 La compositions. For the x = 0 ribbons sample, the surface solidified in contact with the chamber atmosphere (on the side typically called the *free side*) is shown in Figure 4. 10 (a) while the surface which solidified on the copper wheel (the ribbon *wheel side*) is shown in Figure 4. 10 (b, c). The grooves visible on the ribbon wheel side are caused by the "gas pockets" created during melt-spinning by Ar gas from the chamber atmosphere entrapped at imperfect contact between the flowing melt and the rotating copper wheel. The imperfect contact between the wheel and the melt, non-uniform solidification rates are attained on the ribbon surface which generates a non-uniform microstructure, with significant variation in grain size. A much coarser, micron-sized grain structure (Figure 4. 10 (c)) is formed in the "gas pockets" as the result of growth due to slower heat extraction rates compared to the regions that solidified at perfect contact between the melt and the copper wheel surface [155, 51]. A comparison of the ultrafine grained regions on the ribbons wheel side reveals different microstructure evolutions (Figure 4. 10 d, e, f). It can be seen that Ce substitution induces grain growth. At 0.1 Ce fraction, the microstructure is non-uniform, with coarser grain of diameters exceeding 100 nm is observed (Figure 4.10 e) while on the Ce-free sample surface the microstructure is more uniform with grain diameters of roughly 50 nm (Figure 4.10 (d)). La substitution on the other hand has a significant grain refinement effect, the grain diameters going down to 20 - 30 nm at x = 0.1 (Figure 4.10 (f)).

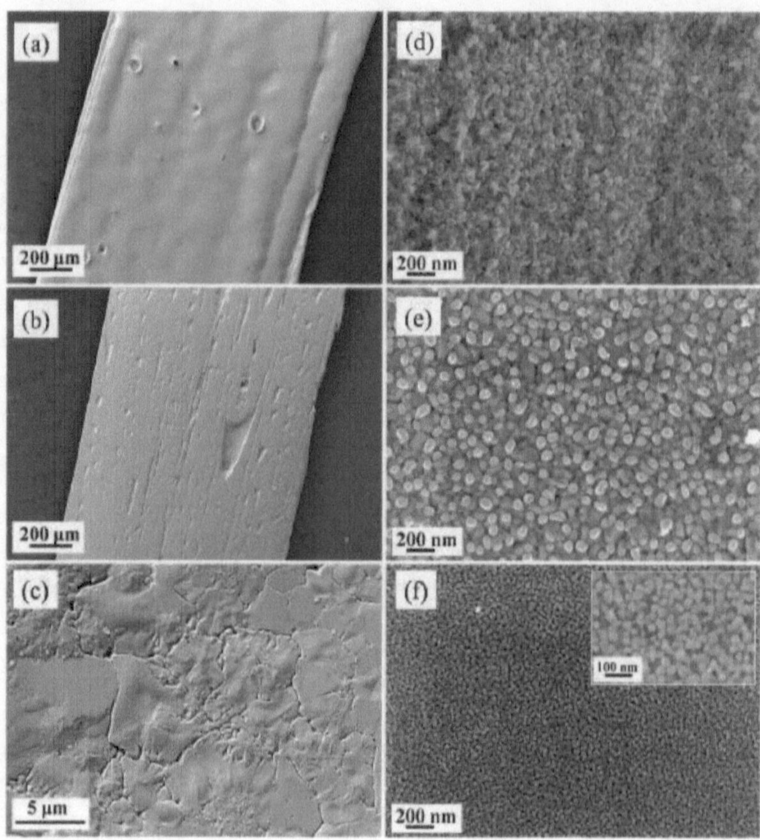

Figure 4.10: SEM images on the free side (a) and wheel side (b) surfaces of the $(Nd_{1-x}R_x)_{13.6}Fe_{73.6}Co_{6.6}Ga_{0.6}B_{5.6}$ melt-spun ribbons at x = 0; coarse grains in the "gas pocket" visible on the wheel side, x = 0 (c); ultrafine microstructure on the wheel side for x = 0 (d), x = 0.1 Ce (e) and x = 0.1 La, with a high magnification detail in the inset picture (f).

The microstructure refinement induced by La substitution was documented in very recent studies on $Ce_2Fe_{14}B$-based alloys [96] and also in older studies on α-Fe/$R_2Fe_{14}B$-type nanocomposites [101, **156**]. The effect could be due to the fact that lanthanum substitution increases the undercooling of the alloy melt (constitutional undercooling) which combined with the local undercooling effect created by very high heat extraction rates [**157**] (through melt chilling at contact with the cold cooper wheel nucleation centers are created) leads to a refined solidified grain structure.

Constitutional undercooling is generated in the alloy by the ejection of La solute atoms from the Φ-phase crystal into the liquid as the solidification front proceeds, progressively enriching the liquid with La and changing its freezing temperature until it ultimately solidifies as the rare-earth-rich eutectic phase. With La solute atom partitioning as described, rapid quenching allows short distance diffusion which leads to grain refinement.

Excessive grain growth can be observed for the $Ce_{13.6}Fe_{73.6}Co_{6.6}Ga_{0.6}B_{5.6}$ melt-spun alloy where microstructure is mainly composed of large, sub-micron grains (Figure 4.11 a, b). At higher La concentrations, a further more pronounced grain size reduction can be observed (Figure 4.11 c, d) with over-quenched areas appearing at the wheel contact surface (Figure 4.11 d). The SEM observations of the La-containing ribbons are in good agreement with the crystallite size reduction suggested by the progressing peak broadening with increasing La concentration on the XRD scans (showing an amorphization tendency at high La content).

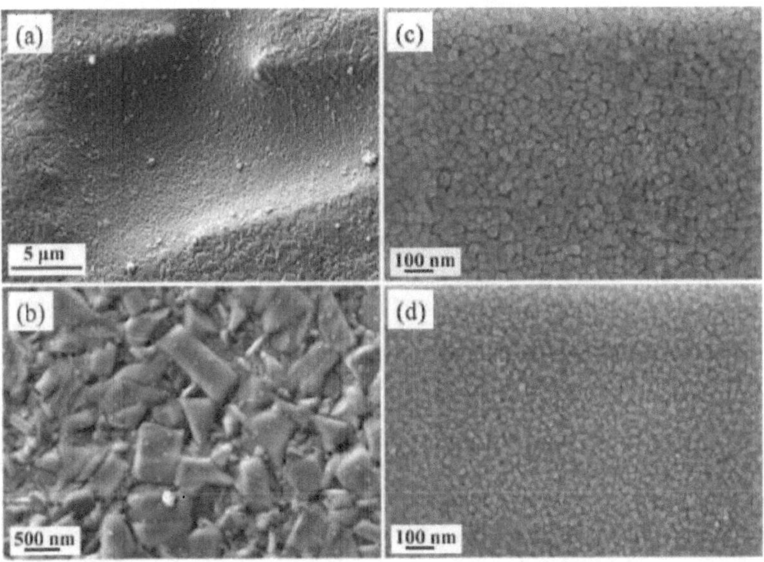

Figure 4.11: SEM images of the $(Nd_{1-x}R_x)_{13.6}Fe_{73.6}Co_{6.6}Ga_{0.6}B_{5.6}$ melt-spun ribbons microstructure on the wheel side surface for compositions with x = 1 Ce fraction (groove - a; wheel contact - b), x = 0.2 La (c) and x = 0.3 La (d) at wheel contact regions.

4.1.3 Hysteresis characteristics and magnetic properties

This section contains an analysis of the magnetic properties of the prepared melt-spun alloys in terms of microstructure and phase composition related hysteresis behavior (magnetization switching with the field and extrinsic magnetic properties) and intrinsic magnetic properties (the saturation polarization determined on the magnetic polarization versus magnetic field curves in the hysteresis loops and Curie transition temperatures derived from magnetization versus temperature curves M(T)).

For all melt-spun powder samples, the magnetic hysteresis measurements were performed at room temperature, up to a maximum applied magnetic field of 8.5 T. The hysteresis loops characteristic to Ce, respectively La-substituted samples $(Nd_{1-x}R_x)_{13.6}Fe_{73.6}Co_{6.6}Ga_{0.6}B_{5.6}$, for x = 0, 0.1, 0.2 and 0.3 fractions, are presented in Figures 4.12 and 4.13. For all Ce-substituted samples, the initial magnetization curves (starting at origin, in the first quadrant) show high susceptibility in low applied magnetic field, a sign of easy domain wall motion and nucleation controlled coercivity mechanism. This effect is the contribution of multi-domain, large grains, the existence of which is confirmed by SEM imaging (seen in Figures 4.10 and 4.11, Section 4.1.2).

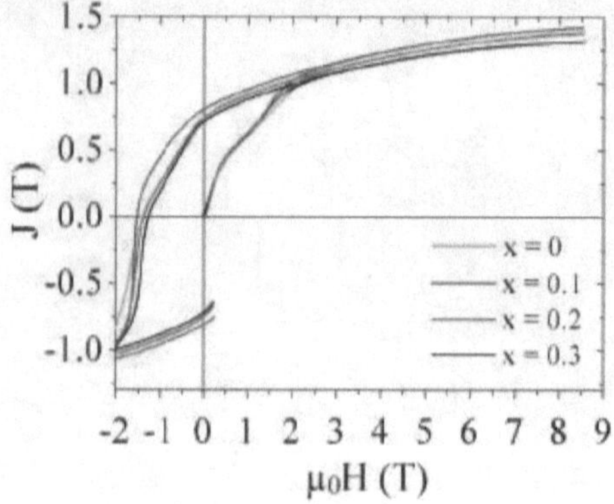

Figure 4.12: Room temperature hysteresis loops of $(Nd_{1-x}Ce_x)_{13.6}Fe_{73.6}Co_{6.6}Ga_{0.6}B_{5.6}$ as-quenched melt-spun alloys where x = 0, 0.1, 0.2 and 0.3.

The hysteresis loops of the $(Nd_{1-x}La_x)_{13.6}Fe_{73.6}Co_{6.6}Ga_{0.6}B_{5.6}$ (x = 0, 0.1, 0.2 and 0.3) melt-spun powders shown in Figure 4.13 indicate an increase in remanence (B_r) up to x = 0.3 La while a severe depression of squareness along with the decrease of coercivity (μ_0H_c) can be seen on the demagnetization curves in the second quadrant.

It is the presence of minor, secondary soft magnetic phases like α-Fe, Nd_2Fe_{17} (for x = 0.5) and over-quenched, amorphous fractions that affects the loop squareness. This explanation is in good agreement with the XRD data showing peaks of the two minor phases and matrix phase peaks broadening (Figure 4.7, Section 4.1.1) and also with the calorimetry measurements that suggest the occurrence of crystallization (Figure 4.16 (b), Section 4.1.4). The initial magnetization curve corresponding to the sample with x = 0.1 La fraction shows low magnetization in low applied field, a sign of domain rotation upon alignment with the applied field which is specific to single-domain particles. It is again a microstructure related effect, the contribution of the finer grain structure obtained by La substitution which generated a more uniformly sized and refined microstructure, diminishing the effects of grain coarsening that occurs on regions with lower solidification rates of the ribbon surface (grain structure in SEM images, Section 4.1.2, Figure 4.9 and 4.10).

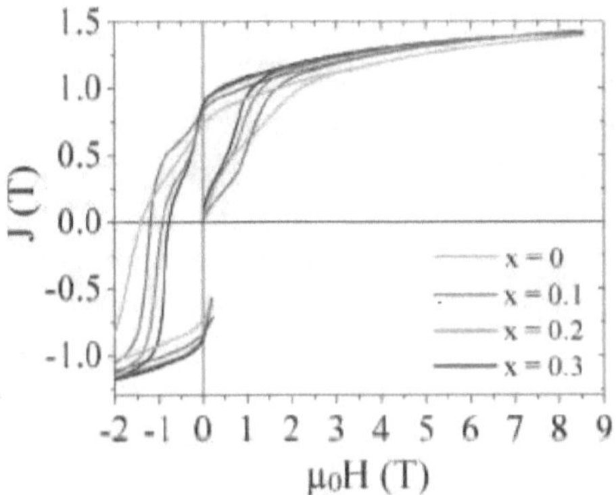

Figure 4.13: Room-temperature hysteresis loops of $(Nd_{1-x}La_x)_{13.6}Fe_{73.6}Co_{6.6}Ga_{0.6}B_{5.6}$ as-quenched melt-spun alloys where x = 0, 0.1, 0.2 and 0.3.

The obtained values for the basic magnetic properties read on the hysteresis curves (remanence B_r, coercivity μ_0H_c and saturation polarization J_s) are shown plotted against Ce and La concentrations in Figure 4.14 (the values are listed together with the T_C points in Tables 4.1 and 4.2). All magnetic properties improve initially with 10 % of the Nd atoms substituted by Ce (Figure 4.14 a-c). After this initial increase (which could be explained by composition variation in the sample material or error), the trends show a strong negative correlation with the Ce concentration, all magnetic properties decreasing with the increasing Ce content in the alloy. Lanthanum substitution (Figure 4.14 d-f) induces a slightly steeper coercivity (μ_0H_c) decrease because of lower anisotropy but, up to the ratio of x = 0.3 La fraction, it maintains saturation polarization (J_s) relatively stable as a result of its tendency to migrate to the grain boundaries rather than enter the $Nd_2Fe_{14}B$ phase.

It enhances the remanence (B_r) which indicates an increased proportion of $Nd_2Fe_{14}B$ phase in these alloys (the formation of $Nd_2Fe_{14}B$ phase is favored since La does not mix with Fe and segregates at the grain boundaries - there is no compound formation in the Fe-La system but the components form an eutectic mixture at 91.5 at. % La concentration [153]). At higher La concentrations, the remanence shows abruptly declining values because of phase segregations. The Curie transition temperatures of the Φ-phase derived from M(T) measurements (the curves shown in Figures 6.5a, 6.5b and 6.6, in the Annex) on the two alloy series are plotted in Figure 4.15. For the Ce-containing alloys T_C shows a slight decrease with increasing Ce concentrations to x = 0.4 (T_{C2} points) and then it plummets as the Ce content is increased. However, in this initial range, the dM/dT curves (Figure 6.5 and 6.6 in the Annex) show additional, lower temperature minima that are situated roughly within 260 - 270 °C range (T_{C1} points in Figure 4.13 a). This is most probably the contribution of a Φ-phase with higher Ce concentration (hence the lower T_C) which separates within this substitution range. In the La-containing samples T_C values for the Φ-phase are only mildly affected and maintained rather high for fractions up to x = 0.4. At x = 0.5 La the Curie temperature shows an increase most probably due to phase segregations that redistribute La outside the Φ-phase. The Curie points remain high as a result of La segregation at the grain boundaries and also due to the lattice expansion produced by the La atoms that are retained within the Φ-phase structure. This argument is further supported by the Curie temperatures constant trend observed for the hydrogenated Φ-phase in $(Nd_{1-x}La_x)_{12.5}Fe_{81.2}B_{6.3}$ (x = 0, 0.1 and 0.2) strip-cast alloys (Curie transition temperatures are indicated on the DTA curves in Figure 6.2 in the Annex section).

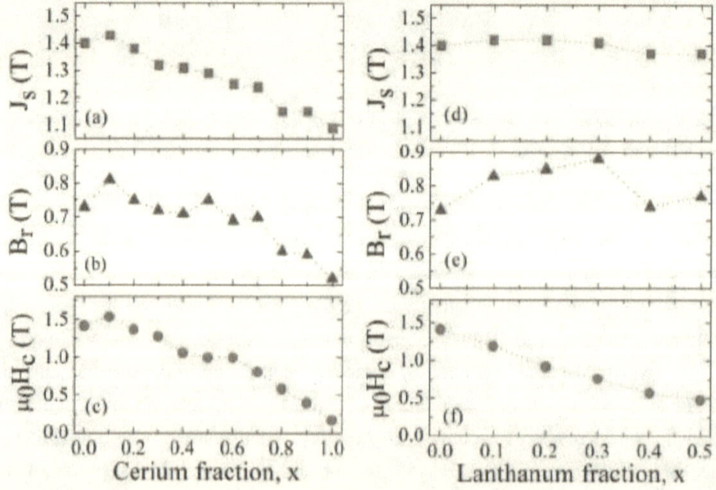

Figure 4.14: Coercivity (μ_0H_c), remanence (B_r) and saturation polarization (J_s) variation with concentration for $(Nd_{1-x}Ce_x)_{13.6}Fe_{73.6}Co_{6.6}Ga_{0.6}B_{5.6}$ (a, b, c) and $(Nd_{1-x}La_x)_{13.6}Fe_{73.6}Co_{6.6}Ga_{0.6}B_{5.6}$ (d, e, f) melt-spun powders.

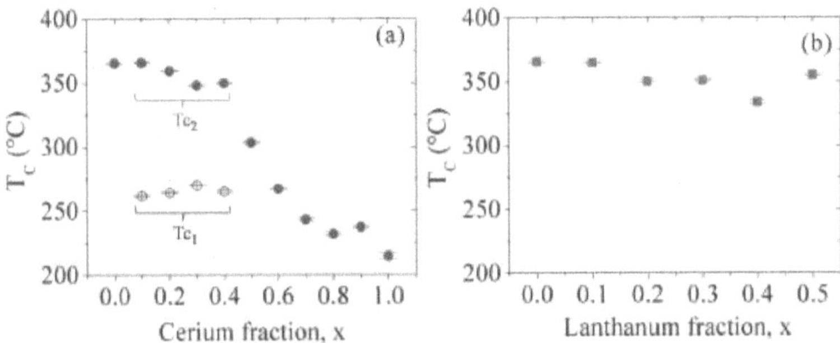

Figure 4.15: Curie temperature of the Φ-phase vs. alloy nominal composition for $(Nd_{1-x}Ce_x)_{13.6}Fe_{73.6}Co_{6.6}Ga_{0.6}B_{5.6}$ (a) and $(Nd_{1-x}La_x)_{13.6}Fe_{73.6}Co_{6.6}Ga_{0.6}B_{5.6}$ (b) melt-spun alloys.

Table 4.1: Magnetic properties (saturation polarization J_s, remanence B_r, coercivity μ_0H_c and Curie transition temperature T_C) of the Ce-substituted melt-spun alloys.

$(Nd_{1-x}Ce_x)_{13.6}Fe_{73.6}Co_{6.6}Ga_{0.6}B_{5.6}$				
x ratio	J_s (T)	B_r (T)	μ_0H_c (T)	T_C (° C)	T_{C1} (° C)
0	1.40	0.74	1.41	365.44 ± 0.7	-
0.1	1.43	0.82	1.51	365.72 ± 0.57	261.87 ± 0.53
0.2	1.38	0.75	1.35	359.27 ± 0.3	264.26 ± 0.45
0.3	1.32	0.72	1.24	347.98 ± 0.6	270.06 ± 0.11
0.4	1.31	0.72	1.04	349.8 ± 0.96	265.35 ± 0.76
0.5	1.29	0.75	0.99	303.38 ± 0.67	-
0.6	1.25	0.70	0.99	267.49 ± 1.06	-
0.7	1.24	0.70	0.81	243.12 ± 1.3	-
0.8	1.15	0.60	0.56	231.81 ± 1.21	-
0.9	1.15	0.60	0.40	237.19 ± 0.75	-
1	1.09	0.54	0.17	214.75 ± 2.39	-

Table 4.2: Magnetic properties (saturation polarization J_s, remanence B_r, coercivity μ_0H_c and Curie transition temperature T_C) of the La-substituted melt-spun alloys.

Alloy	$(Nd_{1-x}La_x)_{13.6}Fe_{73.6}Co_{6.6}Ga_{0.6}B_{5.6}$			
x ratio	J_s (T)	B_r (T)	μ_0H_c (T)	T_C (° C)
0	1.40	0.74	1.41	365.44 ± 0.70
0.1	1.42	0.83	1.19	364.55 ± 0.42
0.2	1.42	0.85	0.92	350.09 ± 0.87
0.3	1.41	0.88	0.76	350.87 ± 0.65
0.4	1.37	0.74	0.57	334.07 ± 1.27
0.5	1.37	0.77	0.48	355.03 ± 1.03

4.1.4 Thermal properties

Figure 4.16 shows DSC traces of the as-spun alloys at different substitution fractions of Ce (measured on selected compositions, (a)) and La (b).

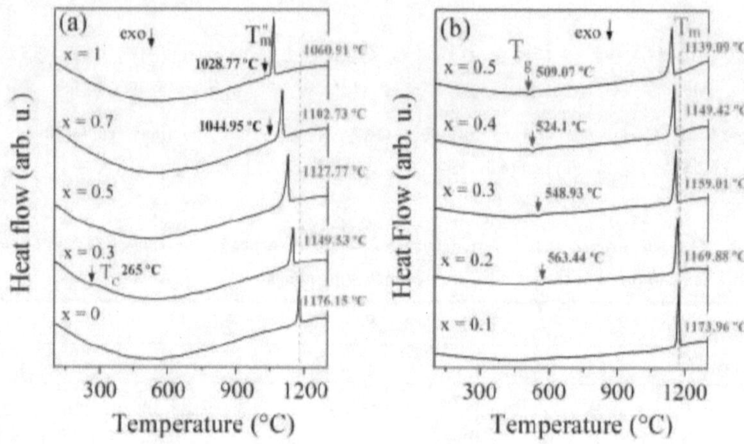

Figure 4.16: DSC scans run at 10 K/min heating hate for $(Nd_{1-x}Ce_x)_{13.6}Fe_{73.6}Co_{6.6}Ga_{0.6}B_{5.6}$ (a) and $(Nd_{1-x}La_x)_{13.6}Fe_{73.6}Co_{6.6}Ga_{0.6}B_{5.6}$ (b) melt-spun powders (T_m = melting temperature of the Φ-phase; T_m'' = secondary phase melting; T_g = crystallization temperature).

On all curves, the strong endothermic effects mark the melting of the Φ-phase. With increasing substitution fractions, both Ce and La progressively shift the melting temperature (T_m) of the Φ-phase to lower temperatures. For equal substitution ratios, the T_m of the Φ-phase is lower for the Ce-containing alloys. The melting point of the rare-earth-rich intergranular phase could not be distinguished on the curves. The weak endothermic peaks in Figure 4.16 (a) correspond to the melting that occurs through the peritectic decomposition of $CeFe_2$ phase (T_m'' situated at 1044.95 °C for x = 0.7 and at 1028.77 °C for x = 1). For the x = 0.3 Ce composition, a Curie transition is suggested by the step at 265 °C which could confirm the presence of the lower T_C phase identified by thermomagnetic analysis. In the case of the alloys with x = 0.2 to x = 0.5 La concentrations, weak exothermic effects mark the crystallization of amorphous fractions which occurs at decreasing temperatures with increasing La content [101]. At x = 0.2 La ratio the unit cell volume expansion begins, showing the retention of La atoms in the Φ-phase structure (crystallite size reduction and amorphization are the associated tendencies).

4.1.5 Hot-worked magnets - microstructure and magnetic properties

Being expected to bring an acceptable trade-off between the price advantage and the magnetic property dilution [158], the melt-spun alloys of composition $(Nd_{1-x}Ce_x)_{13.6}Fe_{73.6}Co_{6.6}Ga_{0.6}B_{5.6}$ with Ce fractions $x = 0.2$ and 0.3 were further used for the preparation of hot-pressed and hot-deformed magnets.

Figure 4.17 shows SEM micrographs from the inner axial fracture surface of the hot-pressed (a, c) and hot-deformed (b, d) magnet of composition $(Nd_{0.7}Ce_{0.3})_{13.6}Fe_{73.6}Co_{6.6}Ga_{0.6}B_{5.6}$. In the hot-pressed state the magnet is composed of nanosized, equiaxed polyhedral grains along with larger, micron-sized grains. The fine grain structure features larger grains comparing to the starting powders suggesting some degree of grain coarsening occurred during hot-pressing. However, the large grains micron-sized grains visible in both the hot-pressed and the hot-deformed body (where they appear undeformed) pre-exist the hot-working processing, originating in the initial non-uniformly solidified structure obtained by melt-spinning. The hot-deformed microstructure features well aligned platelet-like shaped grains.

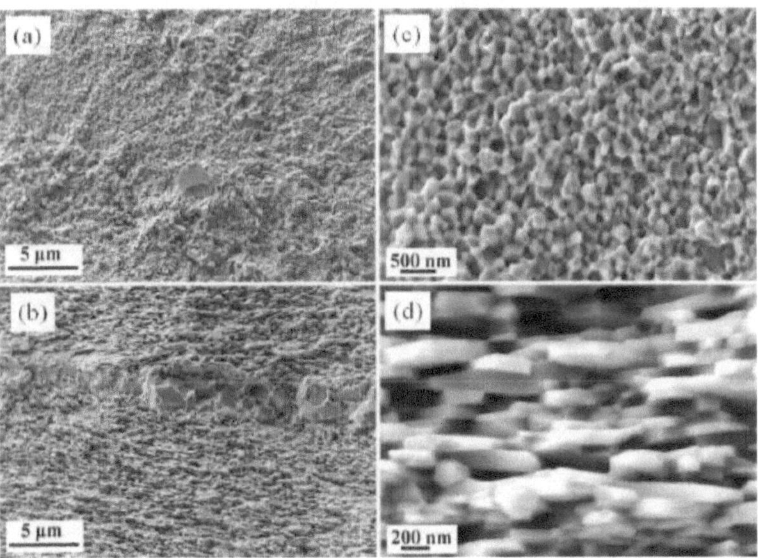

Figure 4.17: Low and high magnification SEM micrographs on cross-section fracture surface of the magnet of $(Nd_{1-x}Ce_x)_{13.6}Fe_{73.6}Co_{6.6}Ga_{0.6}B_{5.6}$ $(x = 0.3)$ composition in hot-pressed (a, c) and hot-deformed (b, d) processing states.

For comparison, the $(Nd_{1-x}La_x)_{13.6}Fe_{73.6}Co_{6.6}Ga_{0.6}B_{5.6}$ melt-spun alloy with x = 0.2 La fraction was also processed by hot-pressing and hot-deformation. The alloy responded well to hot-pressing but suffered extensive cracking during hot-deformation. The sample cleaved transversally and disintegrated into thin slices during preparation for magnetic measurements.

The room temperature demagnetization curves measured of Ce-substituted hot-pressed and hot-deformed magnets are shown in Figure 4.18 and their respective magnetic properties (coercivity μ_0H_c, remanence B_r and energy density product $|BH|_{max}$) are listed in Table 4.3.

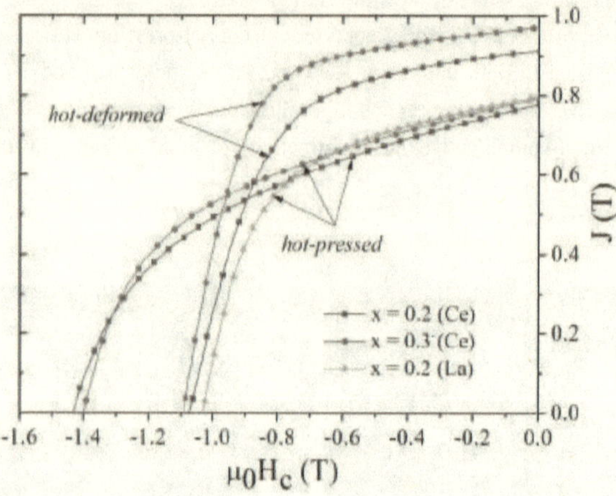

Figure 4.18: Room temperature demagnetization curves of $(Nd_{1-x}Ce_x)_{13.6}Fe_{73.6}Co_{6.6}Ga_{0.6}B_{5.6}$ (x = 0.2 and 0.3) and $(Nd_{1-x}La_x)_{13.6}Fe_{73.6}Co_{6.6}Ga_{0.6}B_{5.6}$ (x = 0.2) hot-worked magnets.

Table 4.3: Magnetic properties (coercivity μ_0H_c, remanence B_r and energy product $|BH|_{max}$) of $(Nd_{1-x}Ce(La)_x)_{13.6}Fe_{73.6}Co_{6.6}Ga_{0.6}B_{5.6}$ (x = 0.2 and 0.3 Ce and x = 0.2 La) magnets after hot-pressing and hot-deformation.

Substitution ratio, x	Hot-pressed			Hot-deformed		
	μ_0H_c (T)	B_r (T)	$|BH|_{max}$ kJ/m³	μ_0H_c (T)	B_r (T)	$|BH|_{max}$ kJ/m³
Ce: x = 0.2	1.43	0.77	100	1.07	0.91	148
Ce: x = 0.3	1.40	0.79	104	1.09	0.97	170
La: x = 0.2	1.03	0.80	107	-	-	-

In hot-pressed state the magnets show similar properties while an increase in remanence is observed for x = 0.3 after hot-deformation. The increased remanence at higher Ce content is due to the improved deformability [159, 160] attributed to the reduction of the intergranular rare-earth-rich phase melting temperature induced by Ce substitution for Nd [161]. The DSC analysis showed decreased melting temperatures of the Φ-phase with increasing Ce concentrations and one may further deduce that the melting temperature of the Nd-rich phase is also decreased according to the eutectic point in the Ce-Fe system which situated at 600 °C [162]. A lower melting temperature of the rare-earth-rich phase implies that the intergranular material flows easily around the grains, enhancing grain deformation and microstructure alignment/texturing.

At the x = 0.2 La concentration, a plausible explanation for the poor deformability would be the presence of lanthanum (raising the melting point of the intergranular phase) and lanthanum oxide at the grain boundaries. Lanthanum also introduces lattice strain (besides crystallite size reduction, strain is also a contribution in the wider XRD peaks breath) which can induce microcracking under pressure load, ultimately producing sample fracture.

4.2   Hydrogen treatment of $(Nd_{1-x}Ce_x)_{15}Fe_{79}B_6$ strip-cast alloys

This subchapter contains an analysis on the effects of Ce substitution in $(Nd_{1-x}Ce_x)_{15}Fe_{79}B_6$ at. % (where x = 0, 0.1, 0.2, ... 0.6) strip-cast alloys processed by hydrogen treatments (hydrogen decrepitation - HD and dynamic hydrogenation-disproportionation-desorption-recombination - d-HDDR). The use of HDDR is proposed here as an alternative processing technique for the fabrication of Ce-substituted $Nd_2Fe_{14}B$-based anisotropic permanent magnets. The transformations occurring in the alloy microstructure along with the changes in phase composition and crystal properties of the Φ-phase are revealed and discussed in correlation with the Ce content and the processing state of the alloys. The influence of Ce substitution on the as-cast microstructure and phase composition of the alloys and the consequent response to hydrogen treatment is analyzed. The variation of magnetic properties of the obtained $(Nd_{1-x}Ce_x)_{15}Fe_{79}B_6$ HDDR powders with respect to Ce concentration is presented and the magnetic property development through HDDR treatment is explained.

The alloys of nominal composition $(Nd_{1-x}Ce_x)_{15}Fe_{79}B_6$ at. % and a measured oxygen content of ~ 300 ppm (for samples with x = 0, 0.2, and 0.3) were prepared by strip-casting at 7 m/s disk speed. The elemental composition of the alloys was determined by ICP-OES measurements for all strip-cast alloys, concentrations corresponding to each composition are listed in Table 4.4. The Al and Si impurities detected in the strip-cast materials originate from the alumina suspension coating applied as on the inside wall of the crucible for protection against the erosion caused by the molten alloy.

Table 4.4: Elemental composition of $(Nd_{1-x}Ce_x)_{15}Fe_{79}B_6$ strip-cast alloys determined by ICP-OES measurements (concentrations are expressed in at. %).

Element	Cerium fraction, x						
	x = 0	x = 0.1	x = 0.2	x = 0.3	x = 0.4	x = 0.5	x = 0.6
B	6.92	7.12	7.05	6.88	7.05	6.97	7.36
Ce	0.00	1.48	2.89	4.30	5.79	7.25	8.69
Fe	77.74	77.77	77.81	78.16	77.93	78.00	77.67
Nd	14.42	12.81	11.48	9.96	8.56	7.22	5.75
Si	0.22	0.18	0.20	0.21	0.17	0.16	0.16
Al	0.69	0.64	0.57	0.50	0.50	0.41	0.37

The obtained strip-cast flake alloys were used in as-cast condition for characterization and further treatment. A homogenization heat treatment in advance of hydrogen processing was not necessary since the flakes were mostly free of primary α-Fe segregations.

At each major processing stage the phase structure and crystal properties of the strip-cast alloys were analyzed by powder XRD while the phase and element distributions were revealed by SEM and WDX/EDX elemental mapping. For the microscopy investigations, cross-sections on the strip-cast flakes were prepared by fine mechanical grinding and polishing followed by ion-milling. The hydrogen processed (decrepitated, disproportionated and fully processed by HDDR) powder particles cross-sections were prepared from powder samples embedded in epoxy resin and polished by the same procedure. Ion-milling is used as a final polishing step because it does not cause deterioration along the grain boundaries on the sample surface as classical mechanical polishing often does resulting in the removal of the grain boundary phase.

4.2.1 Phase composition and microstructure in as-cast state

The XRD scans of $(Nd_{1-x}Ce_x)_{15}Fe_{79}B_6$ (x = 0, 0.1,..., 0.6) strip-cast alloys are shown in Figure 4.19. The patterns exhibit the characteristic peaks of the $Nd_2Fe_{14}B$ matrix (Φ-phase) along with the rare-earth-rich phase and the minor $Nd_{1.1}Fe_4B_4$ (or η-phase). For x = 0 to 0.3 there is indication of minor primary α-Fe segregation (peaks at $\approx 52.34°$), few and fine dendrites having also been seen in SEM in flakes with x = 0.2 and 0.3 Ce ratio (images are not introduced here). At higher Ce ratios, the α-Fe segregations get more intense, appearing strongest at x = 0.5. One may observe that within the studied concentration range, $CeFe_2$ phase segregation starts occurring at x = 0.3 Ce fraction. The relative intensity of the $CeFe_2$ phase peaks increases with the Ce fraction x, indicating increasing proportions of $CeFe_2$ in the strip-cast alloys. It can also be seen that the peaks corresponding to the intergranular Nd-rich phase vanish in the case of the alloys with x = 0.6 Ce fraction.

The unit cell parameters *a* and *c* together with the unit cell volume and the tetragonality factor *c/a* of the Φ-phase are listed in Table 4.5. The values of all unit cell parameters decrease with increasing Ce concentration in the alloy suggesting unit cell contraction and decreasing tetragonality as Ce gradually replaces Nd within the matrix phase structure.

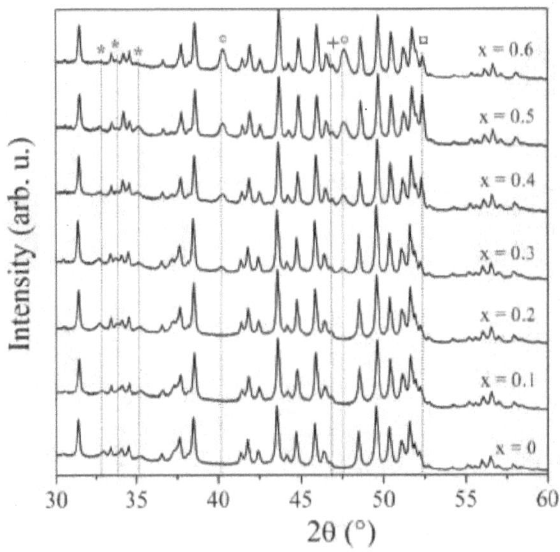

Figure 4.19: X-ray diffraction patterns of the $(Nd_{1-x}Ce_x)_{15}Fe_{79}B_6$ (x – 0, 0.1, ... 0.6) strip-cast alloys. The patterns show the characteristic peaks of the $Nd_2Fe_{14}B$ phase (unmarked), rare-earth-rich phase (*), (Nd, $Ce)_{1.1}Fe_4B_4$ or η-phase (+), α-Fe (□) and $CeFe_2$ (○).

Table 4.5: Values of the Φ-phase unit cell parameters *a* and *c*, volume and tetragonality factor (*c/a*) for x = 0, 0.1 ... 0.6 Ce ratios in $(Nd_{1-x}Ce_x)_{15}Fe_{79}B_6$ strip-cast alloys.

Ce ratio	x = 0	x = 0.1	x = 0.2	x = 0.3	x = 0.4	x = 0.5	x = 0.6
c (Å)	12.2081	12.2024	12.1924	12.1845	12.1786	12.1718	12.1581
a (Å)	8.8050	8.8019	8.7991	8.7945	8.7907	8.7863	8.7798
c/a	1.3865	1.3863	1.3857	1.3855	1.3854	1.3853	1.3848
V_{cell} (Å³)	946.46	945.37	943.98	942.38	941.13	939.65	937.20

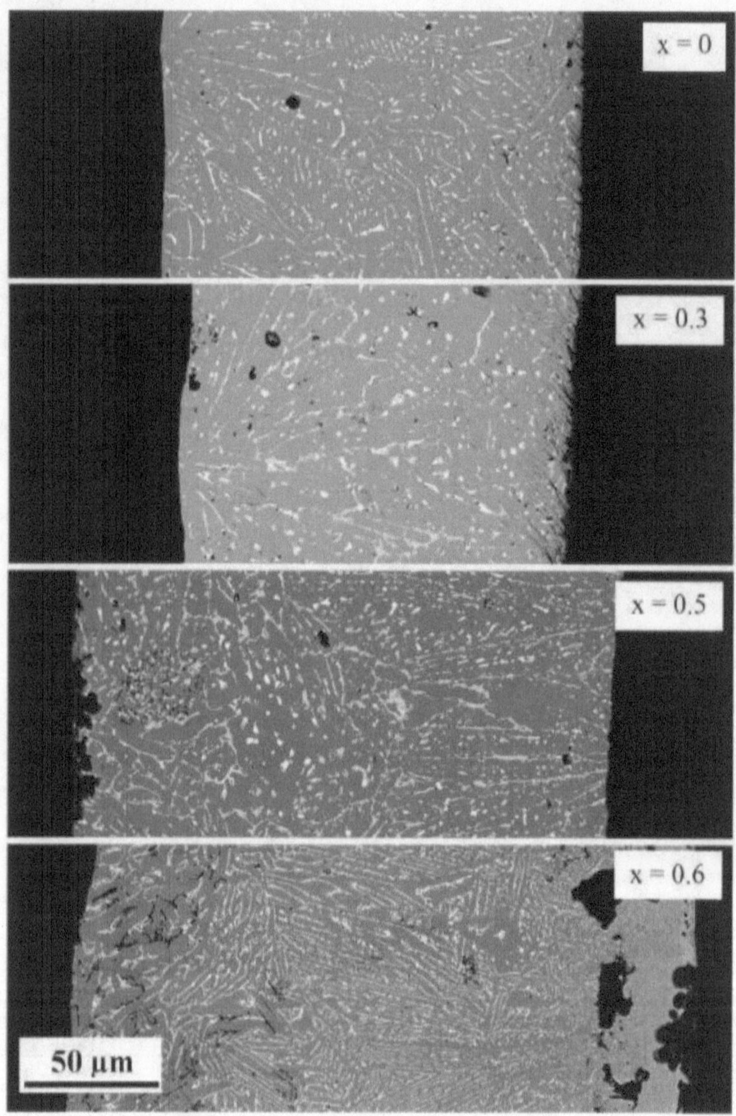

Figure 4.20: BSE micrographs of the ion-polished cross-section surface of $(Nd_{1-x}Ce_x)_{15}Fe_{79}B_6$ strip-cast flakes for x = 0, 0.3, 0.5 and 0.6 Ce concentrations.

Figure 4.20 shows the microstructure on the ion-polished cross-section surface of the strip-cast flakes samples with corresponding x = 0, 0.3, 0.5 and 0.6 Ce concentrations. The obtained strip-cast flakes have a varying thickness of roughly 150 to 300 µm. The flakes feature the typical strip-cast microstructure with lamellar morphology of the matrix phase grains which appear in the micrographs in dark gray contrast. The intergranular material is visible in lighter contrast. In the Ce-free sample the rare-earth-rich intergranular phase appears in white contrast (because of the fine grain structure the η-phase also located at the intergranular regions is not visible in this magnification). The $CeFe_2$ phase also lying at the intergranular regions shows in light gray contrast. The images confirm the XRD data (shown in Figure 4.19) and reveal that $CeFe_2$ replaces the Nd-rich phase at the grain boundaries, becomes predominant in the alloys with high Ce content. In the alloy with x = 0.6 Ce fraction the $CeFe_2$ phase completely surrounds the matrix phase grains. The EPMA-WDX elemental mappings performed on a magnified surface detail of the Ce-free strip-cast flake cross-section presented in Figure 4.21. They show Nd concentrated at the grain boundaries constituting the rare-earth-rich phase. Also the B mapping reveals spots of higher concentrations indicating the presence of the η-phase boride laying within the intergranular material.

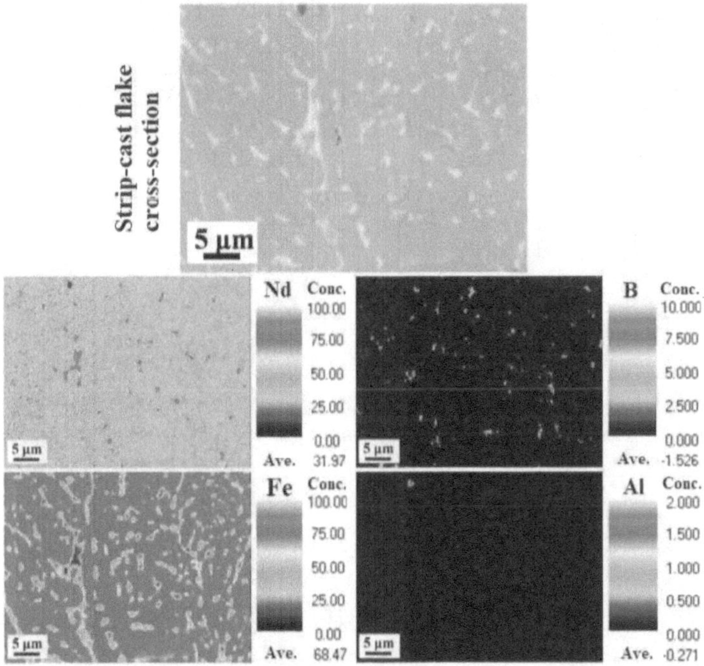

Figure 4.21: BSE micrograph and its corresponding quantified WDX-EPMA elemental maps (color scale in wt. %) on the ion-polished cross-section surface of a strip-cast flake with $Nd_{15}Fe_{79}B_6$ at. % ($Nd_{32.58}Fe_{66.44}B_{0.98}$ wt. %) nominal composition.

Figure 4.22 shows a BSE micrograph with associated WDX-EPMA elemental mappings on the cross-section surface of a strip-cast flake sample with x = 0.2 Ce concentration. One may observe that Ce is mainly concentrated at the grain boundary regions, mixed within the Nd-rich phase. The concentrations of Al quantified on the elemental mappings of these strip-cast flakes do not reflect the overall concentration of the element in the alloy (checked by ICP-OES, results are shown in Table 4.4).

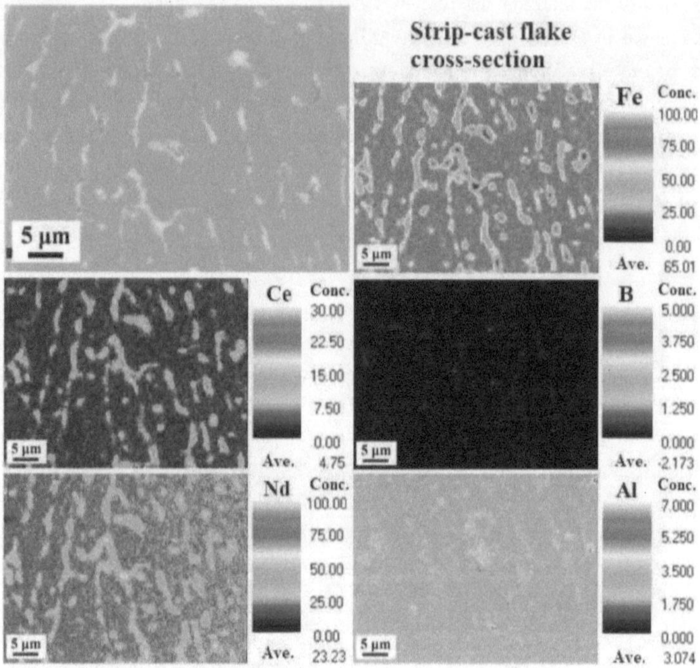

Figure 4.22: BSE micrograph and its corresponding quantified WDX-EPMA elemental maps (color scale in wt. %) on the ion-polished cross-section surface of a strip-cast flake with nominal composition $(Nd_{1-x}Ce_x)_{15}Fe_{79}B_6$ for x = 0.2 $(Nd_{26.11}Ce_{6.34}Fe_{66.57}B_{0.98}$ wt. %).

Figure 4.23 presents the BSE image and the corresponding WDX-EPMA mappings of a high magnification detail on the cross-section surface of a strip-cast flake sample with x = 0.3 Ce concentration. It shows Ce concentrated at the grain boundary within the rare-earth-rich phase, mixed in the η-boride which is clearly visible now (darker gray contrast within the intergranular regions on the BSE image) and constituting the $CeFe_2$ phase (light gray contrast at the grain boundaries).

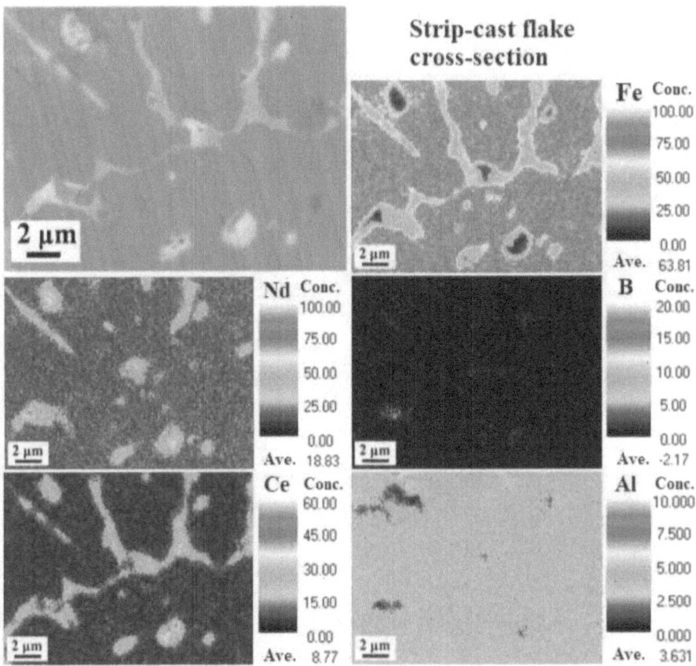

Figure 4.23: BSE micrograph and its corresponding quantified WDX-EPMA elemental maps (color scale in wt. %) on the ion-polished cross-section surface of a strip-cast flake with nominal composition $(Nd_{1-x}Ce_x)_{15}Fe_{79}B_6$ for $x = 0.3$ ($Nd_{22.87}Ce_{9.52}Fe_{66.63}B_{0.98}$ wt. %).

The BSE image and the corresponding EDX mappings of a magnified region on the cross-section surface of a strip-cast flake sample with $x = 0.6$ Ce concentration is presented in Figure 4.24. It shows $CeFe_2$ in light gray forming the intragranular material at the edges of the matrix phase grains and the minor η-phase in dark gray contrast.

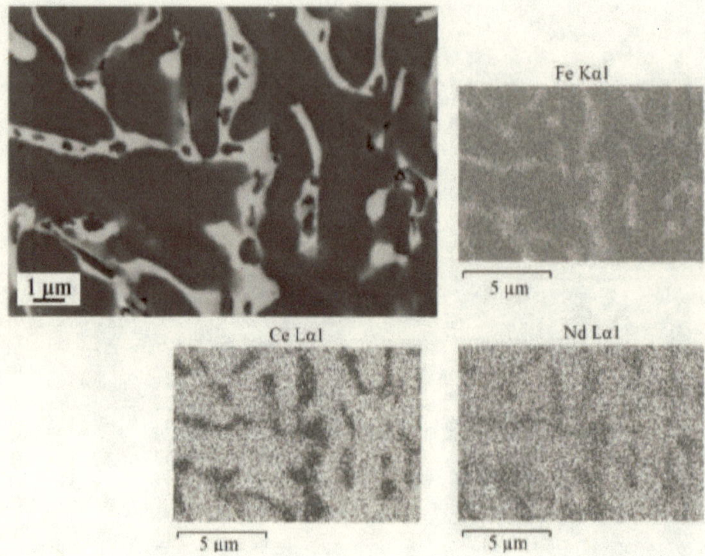

Figure 4.24: BSE image and corresponding EDX elemental mappings of a magnified detail on the cross-section surface of $(Nd_{1-x}Ce_x)_{15}Fe_{79}B_6$ strip-cast flake at x = 0.6 Ce concentration.

4.2.2 Phase composition and microstructure in decrepitated state

The XRD patterns in Figure 4.25 lack the X-ray signature of the $CeFe_2$ phase. $CeFe_2$ showed peaks at $2\theta = 40.26°$ and $47.61°$ for the strip-cast alloys (shown in Figure 4.18, section 4.2.1) and is no longer detected for the decrepitated powder alloys. Studies have shown that H_2 induces amorphization in Laves-type compounds including $CeFe_2$ which exothermically absorbs H_2 and transforms to $CeFe_2H_x$ in amorphous state [163] with reaction temperatures rising to ~50 °C [67]. The η-phase peaks are now well differentiated as the $Nd_2Fe_{14}B$ hydrogenated phase peaks are shifted to lower angles (the lattice is expanded by H_2 absorbed interstitially). Additionally, the peaks of the hydrogenated Φ-phase shift to higher angles with increasing Ce substitution ratios of the alloys suggesting unit cell contraction with increasing Ce concentration, in good agreement with the existing literature which reported larger lattice parameters for $Nd_2Fe_{14}BH_x$ in comparison to $Ce_2Fe_{14}BH_x$ regardless of stoichiometry [39, 40]. One can observe that also the η-phase peaks shift to higher angles with increasing Ce concentration which indicates lattice contraction as Ce, which has smaller atomic radius, substitutes Nd in its structure.

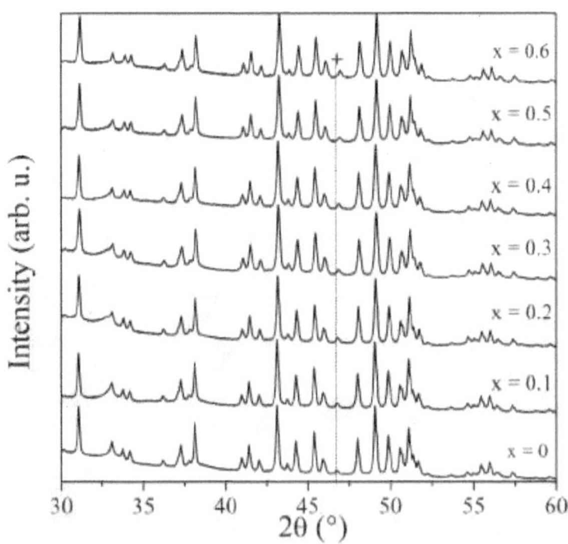

Figure 4.25: X-ray diffraction patterns of the $(Nd_{1-x}Ce_x)_{15}Fe_{79}B_6$ (x = 0, 0.1 ... 0.6) decrepitated alloys showing peaks of $Nd_2Fe_{14}BH_x$ (unmarked) and $Nd_{1.1}Fe_4B_4$ (η-phase).

Figures 4.26, 4.27 and 4.28 contain BSE images with corresponding WDX-EPMA mappings of ion-polished surfaces of decrepitated powder samples of $(Nd_{1-x}Ce_x)_{15}Fe_{79}B_6$ composition for x = 0, 0.2 and 0.3 respectively. Besides phase contrast, the BSE images reveal intergranular and intragranular cracks on the powder surface. The material fractured along the grain boundaries and also through the grains due to the large difference in volume expansion upon hydrogenation between the $Nd_2Fe_{14}B$ phase and the rare-earth-rich phase.

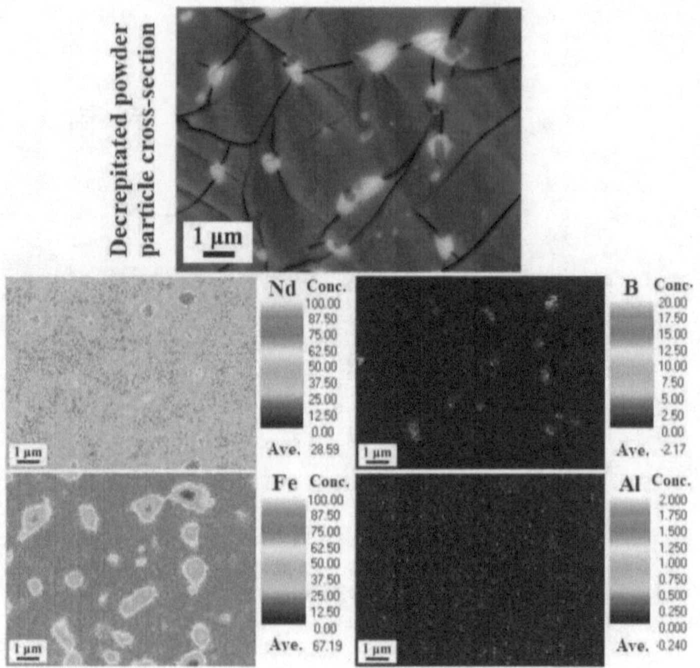

Figure 4.26: BSE image and corresponding WDX-EPMA elemental maps (color scale in wt. %) on the cross-section surface of a hydrogen decrepitated powder particle obtained from the strip-cast alloy of $Nd_{15}Fe_{79}B_6$ at. % ($Nd_{32.58}Fe_{66.44}B_{0.98}$ wt. %) nominal composition.

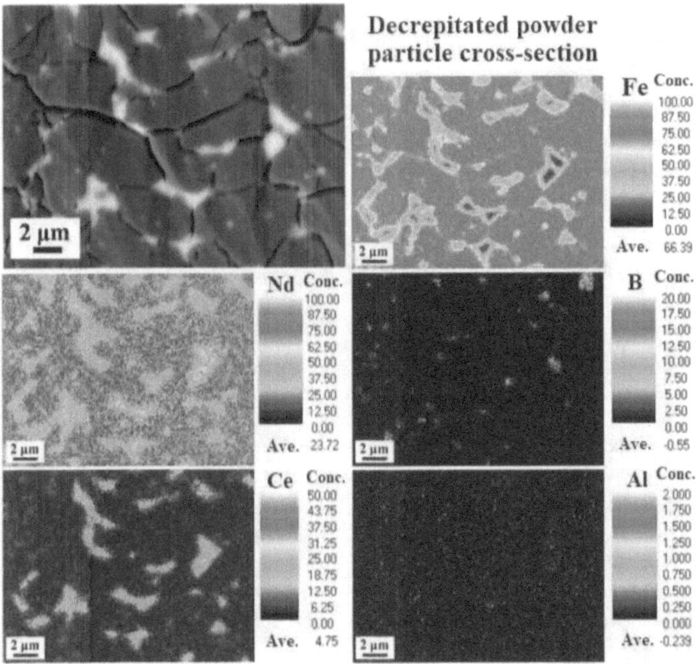

Figure 4.27: BSE image and corresponding quantified WDX-EPMA elemental maps (color scale in wt. %) on the cross-section surface of a hydrogen decrepitated powder particle obtained from the strip-cast alloy of $(Nd_{1-x}Ce_x)_{15}Fe_{79}B_6$ at. % (x = 0.2) or $Nd_{26.11}Ce_{6.34}Fe_{66.57}B_{0.98}$ wt. % nominal composition.

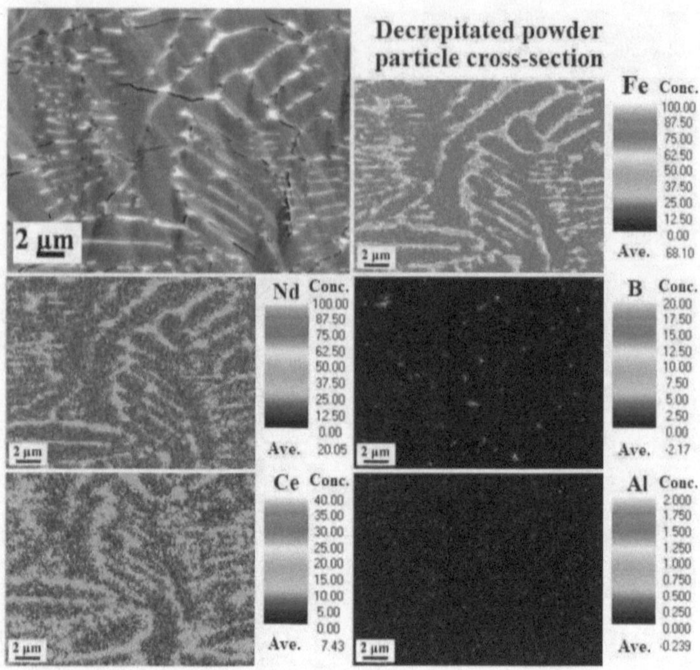

Figure 4.28: BSE image and the corresponding quantified WDX-EPMA elemental maps (color scale in wt. %) on the cross-section surface of a hydrogen decrepitated powder particle obtained from the strip-cast alloy of $(Nd_{1-x}Ce_x)_{15}Fe_{79}B_6$ at. % (x = 0.3) or $Nd_{22.87}Ce_{9.52}Fe_{66.63}B_{0.98}$ wt. % nominal composition.

4.2.3 Phase composition and microstructure in disproportionated state

XRD patterns in Figure 4.29 show the peaks of α-Fe, $(Nd, Ce)H_x$ and Fe_2B. All alloys are in fully disproportionated condition which is confirmed by the absence of the $Nd_2Fe_{14}B$ phase characteristic reflections. The diffraction peaks of the rare-earth hydride $(Nd, Ce)H_x$ resulted from the disproportionation reaction shift to lower angles with increasing Ce concentration in the alloy. This suggests lattice expansion in the face centered cubic (fcc) hydride structure with increasing Ce ratio and is explained by the larger lattice parameter of CeH_x comparing to NdH_x [164].

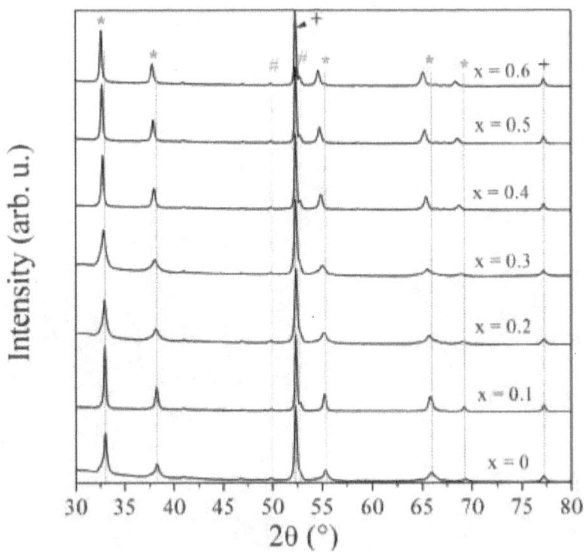

Figure 4.29: XRD patterns of the disproportionated alloys $(Nd_{1-x}Ce_x)_{15}Fe_{79}B_6$ (x = 0, 0.1, ... 0.6) showing peaks of the (Nd, Ce)H_x (*), α-Fe (+) and Fe$_2$B (#) constituent phases.

Figures 4.30, 4.31 and 4.32 present the BSE images and the related WDX-EPMA mappings of ion-polished surfaces of hydrogen disproportionated powder samples of $(Nd_{1-x}Ce_x)_{15}Fe_{79}B_6$ nominal composition for x = 0, 0.2 and 0.3 respectively.

The BSE images reveal the hydrogen disproportionated microstructure featuring the typical disproportionation reaction products forming within the confines of the initial cast grain structure. It consists of the finely divided mixture of α-Fe (gray contrast), Fe$_2$B in black contrast (not easily distinguishable here but clearly shown later in Figure 4.36 in Section 4.2.4) and NdH$_x$ (or Nd(Ce)H$_x$ where x = 0.2 and 0.3) in bright contrast.

The η-phase present at the intergranular region remained unreacted during disproportionation. It is known to partially disproportionate in high pressure H$_2$ at very long exposure time and then partially recombine in vacuum [165].

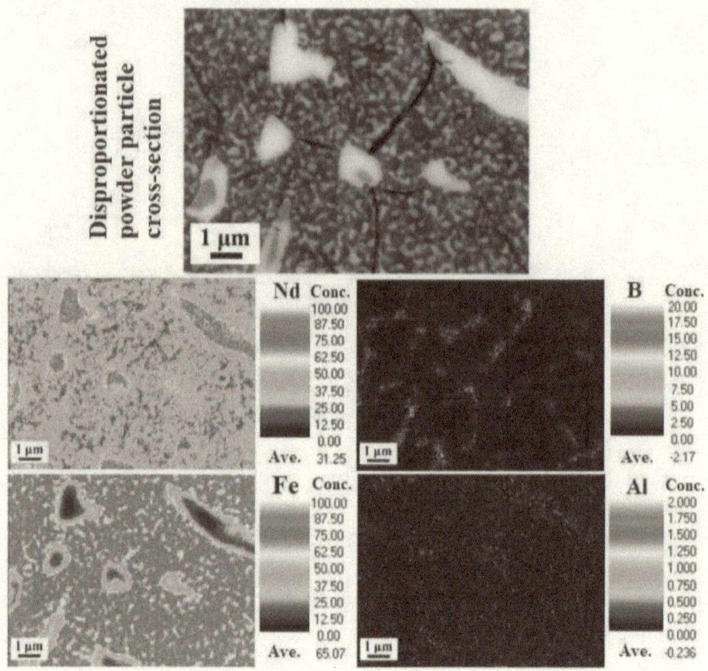

Figure 4.30: BSE image and corresponding WDX-EPMA elemental maps (color scale in wt. %) on the cross-section surface of a hydrogen disproportionated powder particle obtained from the strip-cast alloy of $Nd_{15}Fe_{79}B_6$ at. % ($Nd_{32.58}Fe_{66.44}B_{0.98}$ wt. %) nominal composition.

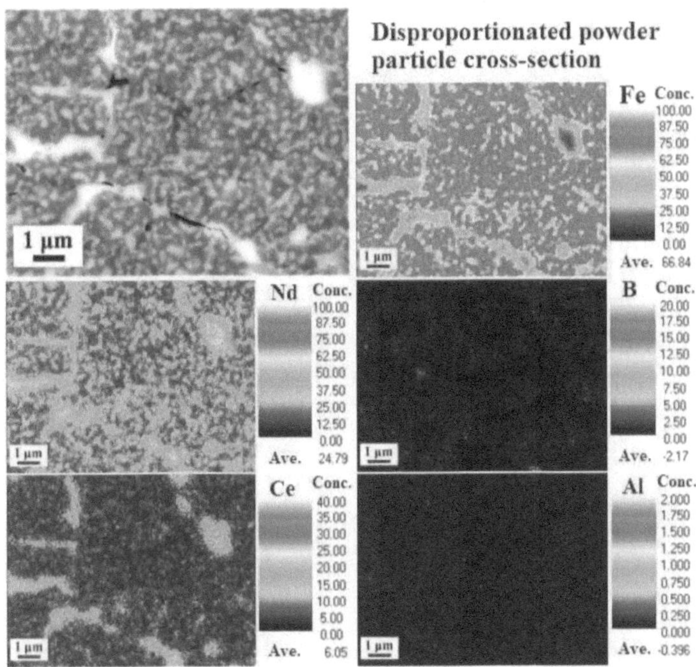

Figure 4.31: BSE image and corresponding quantified WDX-EPMA elemental maps (color scale in wt. %) on the cross-section surface of a hydrogen disproportionated powder particle obtained from the strip-cast alloy of $(Nd_{1-x}Ce_x)_{15}Fe_{79}B_6$ at. % (x = 0.2) or ($Nd_{26.11}Ce_{6.34}Fe_{66.57}B_{0.98}$ wt. %) nominal composition.

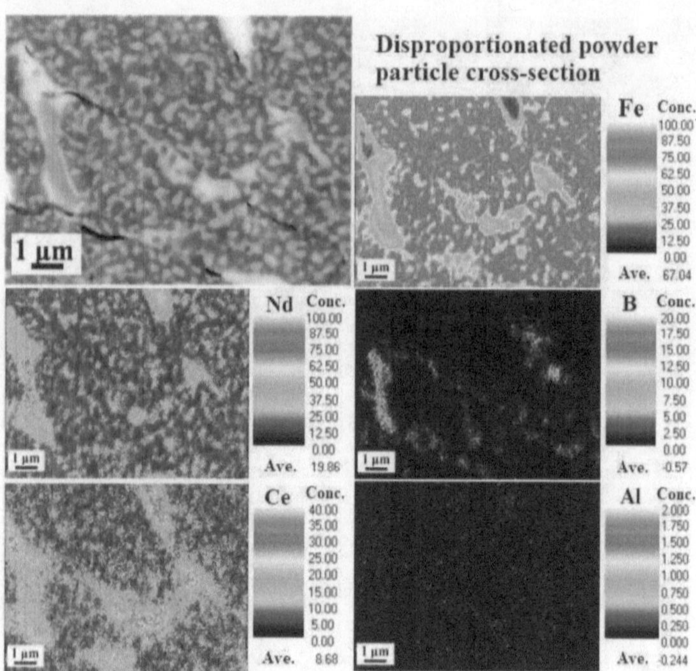

Figure 4.32: BSE image and corresponding quantified WDX-EPMA elemental maps (color scale in wt. %) on the cross-section surface of the hydrogen disproportionated powder particle obtained from the strip-cast alloy of $(Nd_{1-x}Ce_x)_{15}Fe_{79}B_6$ at. % (x = 0.3) or $(Nd_{22.87}Ce_{9.52}Fe_{66.63}B_{0.98}$ wt. %) nominal composition.

The microstructure analysis of the disproportionated powders reveals that the $CeFe_2$ intergranular phase regions (turned into amorphous $CeFe_2H_x$ after H_2 decrepitation) are now found in the form of a separated mixture (Figure 4.33).

The $CeFe_2H_x$ decomposition into the CeH_x and α-Fe mixture is more easily discernible in the samples with higher Ce content where larger amounts of $CeFe_2$ were formed at the grain boundaries in the strip-cast microstructure. The CeH_x and α-Fe mixture is clearly identified in the highlighted region on the BSE image of sample with x = 0.4 Ce substitution ratio (Figure 4.34). Here α-Fe appears in gray contrast surrounded by CeH_x in light gray. The associated EDX maps in the figure reveal the corresponding elemental distribution.

In analogy with the cast microstructure, the micrographs showing the ion-polished powder surface for the disproportionated alloy with x = 0.6 Ce fraction (Figure 4.35) reveal the disproportionated matrix phase grains surrounded by the CeH_x and α-Fe mixture formed by the decomposition of the hydrogenated $CeFe_2$ intergranular phase.

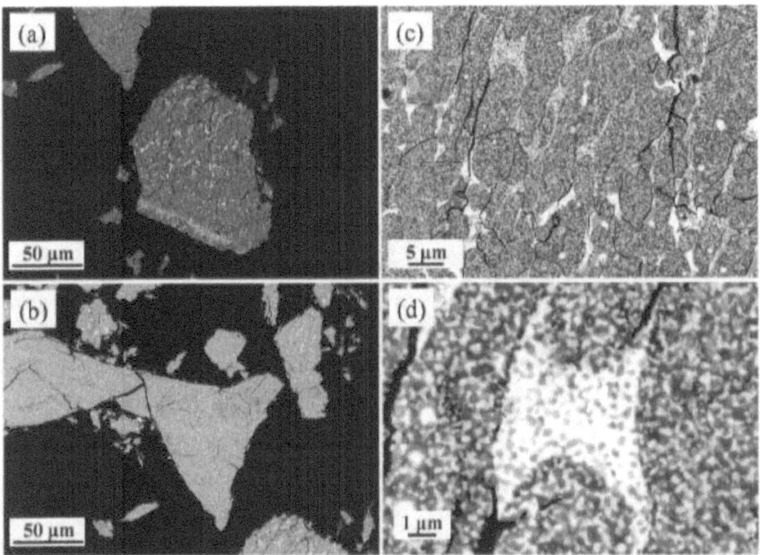

Figure 4.33: BSE micrographs - overview of the surface of embedded and ion-polished powder particles (a) and magnified details (b, c, d) for the $(Nd_{1-x}Ce_x)_{13.6}Fe_{73.6}Co_{6.6}Ga_{0.6}B_{5.6}$ (x = 0.4) disproportionated alloy.

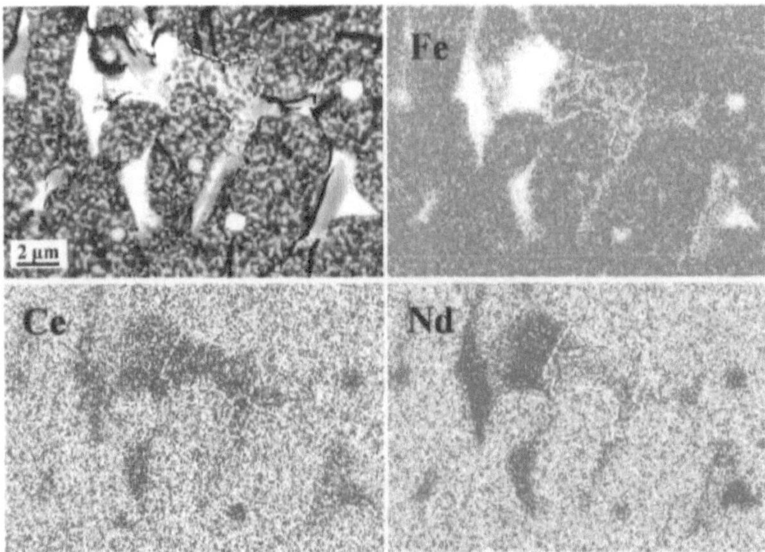

Figure 4.34: BSE image and the related EDX elemental distribution maps of the $(Nd_{1-x}Ce_x)_{15}Fe_{79}B_6$ (x = 0.4 Ce concentration) disproportionated sample.

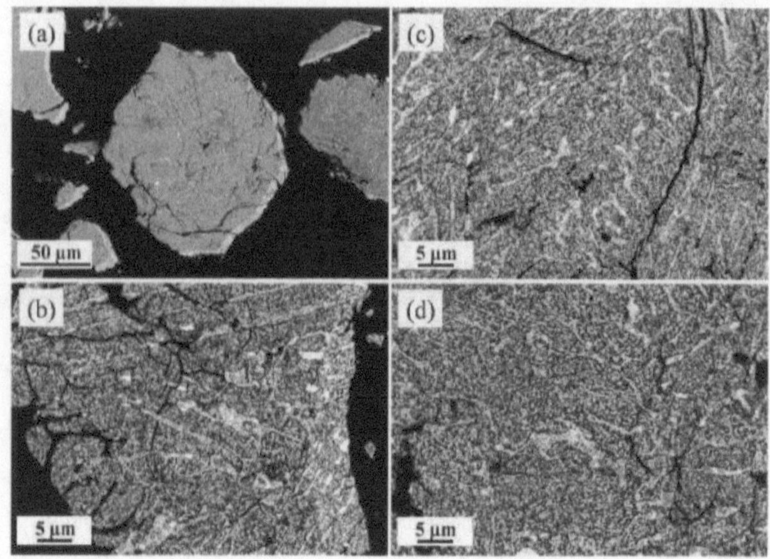

Figure 4.35: BSE micrographs - overview of the surface of embedded and ion-polished powder particles (a) and magnified regions (b, c, d) for the $(Nd_{1-x}Ce_x)_{15}Fe_{79}B_6$ (x = 0.6) disproportionated alloy.

4.2.4 Phase composition and microstructure in recombined state

An overview of the microstructure transformations that occur at the major processing stages is given in Figure 4.36 which presents SEM micrographs of the $(Nd_{1-x}Ce_x)_{15}Fe_{79}B_6$ (x = 0 and 0.3) in as-cast, thermally disproportionated in hydrogen and fully HDDR processed (recombined) conditions.

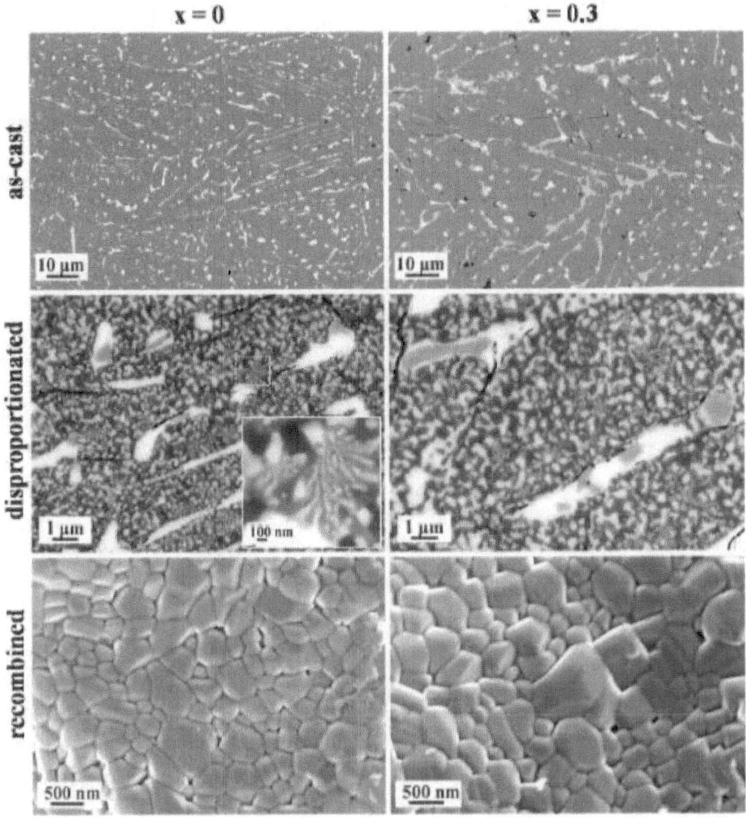

Figure 4.36: SEM micrographs of $(Nd_{1-x}Ce_x)_{15}Fe_{79}B_6$ alloys for x = 0 and 0.3 Ce concentrations, in as-cast (BSE mode, ion-polished flake cross-section surface), disproportionated (BSE mode, resin-embedded and ion-polished powder) and recombined state (SE mode, free powder).

In as-cast (as well decrepitated) and disproportionated conditions the Nd/Ce ratios within the Φ-phase matrix remain the same. The decrepitation process does not alter the concentrations of rare-earth elements in the matrix. It is also safe to assume that in disproportionated state the Nd/Ce ratio within the disproportionated matrix grains remains unchanged. Point concentration measurements were performed on the Φ-phase matrix grains for the as-cast and disproportionated alloys with x = 0 and 0.3 Ce concentrations. The averaged elemental concentrations are listed in the Table 4.6.

Table 4.6: Elemental concentrations (at. %) obtained by EPMA-WDX point analysis performed within the strip-cast and disproportionated matrix phase grains for the $(Nd_{1-x}Ce_x)_{15}Fe_{79}B_6$ alloys with x = 0 and 0.3. The point measurement locations are marked on the corresponding BSE images in the Annex, Figures 6.7, 6.8, 6.9 and 6.10.

Processing state	Cerium fraction, x	Nd	Ce	Fe	B	O	Al	Total
As-cast	0	11.45	0.01	84.12	3.42	0.86	0.14	100
	0.3	6.36	2.45	66.22	0.59	12.50	11.88	100
Disproportionated	0	12.14	0.02	82.06	3.56	1.96	0.27	100
	0.3	8.49	3.27	85.15	1.37	1.57	0.15	100

In the case of the Ce-free alloy, the point analysis quantifications show similar Nd concentration in the as-cast and disproportionated state. Also in the $(Nd_{0.7}Ce_{0.3})_{15}Fe_{79}B_6$ alloy, the Ce/Nd atomic ratio in the Φ-phase is 0.38 in both the as-cast and the disproportionated samples.

Most probably, the Nd/Ce ratios remain the same also in the alloys with higher Ce concentrations (at $x \geq 0.4$). All system components (including the rare-earth hydrides at the grain boundary and within the disproportionated grain) are solid at the disproportionation stage. In this context, since the disproportionation reaction is locally confined within the as-cast grain, significant diffusion of the rare earth elements outside the Φ-phase that would alter the Nd/Ce ratio is unlikely. However, later in the HDDR process, at the desorption-recombination stage, changes in the rare earth elements distribution are expected to occur due to enhanced diffusion by the melting of the intergranular rare-earth components after hydrogen desorption (from the rare-earth-rich hydrogenated phase and CeH_x resulted from $CeFe_2H_x$ decomposition).

Figure 4.37 shows the XRD patterns of the $(Nd_{1-x}Ce_x)_{15}Fe_{79}B_6$ HDDR alloys. $CeFe_2$ peaks show decreased relative intensities after the HDDR process compared to the strip-cast alloys indicating that its fraction is slightly diminished after the treatment. A weak reflection of the Nd-rich phase can now be seen for x = 0.6 while residual Fe_2B is detected in all samples. At this stage, the α-Fe segregations observed in the strip-cast flakes are reduced to minor content (that is for the alloys with notable initial content) or completely eliminated due to the known homogenization effect of the HDDR treatment [71].

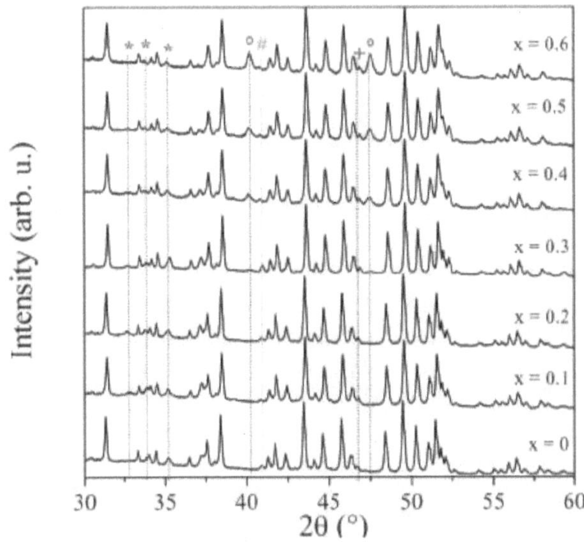

Figure 4.37: X-ray diffraction patterns of the $(Nd_{1-x}Ce_x)_{15}Fe_{79}B_6$ (x = 0, 0.1, … 0.6) fully HDDR processed powders. The patterns show the characteristic peaks of the $Nd_2Fe_{14}B$ phase (not labeled), the rare-earth-rich phase (*), η-phase (+), $CeFe_2$ (○) and Fe_2B (#).

Table 4.7: Values of the Φ-phase unit cell parameters a and c, volume and tetragonality factor (c/a) for x = 0, 0.1 … 0.6 Ce ratios in $(Nd_{1-x}Ce_x)_{15}Fe_{79}B_6$ HDDR alloys.

Ce ratio	x = 0	x = 0.1	x = 0.2	x = 0.3	x = 0.4	x = 0.5	x = 0.6
c (Å)	12.2046	12.2045	12.1965	12.1873	12.1758	12.1729	12.1679
a (Å)	8.7998	8.8025	8.7971	8.7926	8.7883	8.7875	8.7850
c/a	1.3869	1.3865	1.3864	1.3861	1.3855	1.3852	1.3851
V_{cell} (Å³)	945.07	945.65	943.88	942.19	940.38	939.99	939.07

In order to illustrate the evolution of the Φ-phase lattice constants through the process, the unit cell parameters a and c, the unit cell volume and the tetragonality factor c/a are plotted against Ce nominal concentration of the strip-cast and HDDR alloys (Figure 4.38). For both strip-cast (SC) and HDDR alloys the c lattice parameter shows a monotonic decreasing trend with increasing Ce nominal content. However, noticeable increases of the c parameter are observed after HDDR treatment for the Ce concentrations x = 0.1, 0.2, 0.3, 0.5 and 0.6.

The values of a parameter decrease monotonically with increasing x in the case of the as-cast samples and given the c parameter trend, the unit cell volume follows the same pattern.

In comparison to the as-cast condition, after HDDR, the *a* values are lower at x = 0, 0.2, 0.3, and 0.4. The variation of *a* parameter with Ce concentration changes in the HDDR samples, showing an increase at x = 0.1 then decreasing with increasing Ce content. The Ce-free sample shows decreased cell volume and higher tetragonality after HDDR. Volume expansion due to larger *a* and lower tetragonality are observed at x = 0.1. Particularly pronounced at x = 0.2 and 0.3 Ce concentrations, although the cell volume contraction trend applies for both strip-cast and HDDR samples, comparing to the as-cast condition, anisotropic lattice expansion (preferentially along the *c* direction) is observed after the HDDR treatment of these alloys. The *c* parameter and *c/a* factor increase after HDDR indicate that Ce migrates out the Φ-phase structure to the intergranular material during the treatment. Given the lattice constants change from as-cast to HDDR, one may interpret that, discernable especially at 0.2 and 0.3, Ce is expelled to some degree from the Φ-phase and migrates to the grain boundaries at the desorption-recombination stage of HDDR.

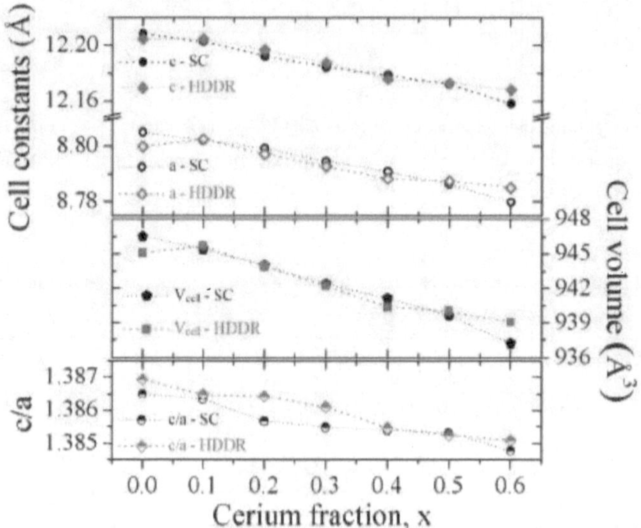

Figure 4.38: Plot of the unit cell parameters *c* and *a*, volume (V_{cell}) and tetragonality factor (*c/a*) of the Φ-phase against Ce nominal content for the $(Nd_{1-x}Ce_x)_{15}Fe_{79}B_6$ (x = 0, 0.1, ..., 0.6) alloys in strip-cast (SC) and HDDR processing states.

SEM images of the $(Nd_{1-x}Ce_x)_{15}Fe_{79}B_6$ HDDR processed materials for x = 0.3 and x = 0.6 are presented in Figure 4.39. At x = 0.3 the sample features a uniformly distributed rare-earth-rich intergranular phase (light gray) around the $Nd_2Fe_{14}B$ grains (dark gray contrast). The η-phase grains (dark contrast) and Nd-enriched regions are visible in bright contrast in the low magnification image in Figure 4.39 (a).

At x = 0.6 composition the intergranular material consists of $CeFe_2$ which is discontinuous among the $Nd_2Fe_{14}B$ recombined grains as suggested by the lack of light contrast at the grain boundaries in the BSE images and the sharp, tightly closed grain edges in the SE image in Figure 4.39 (f).

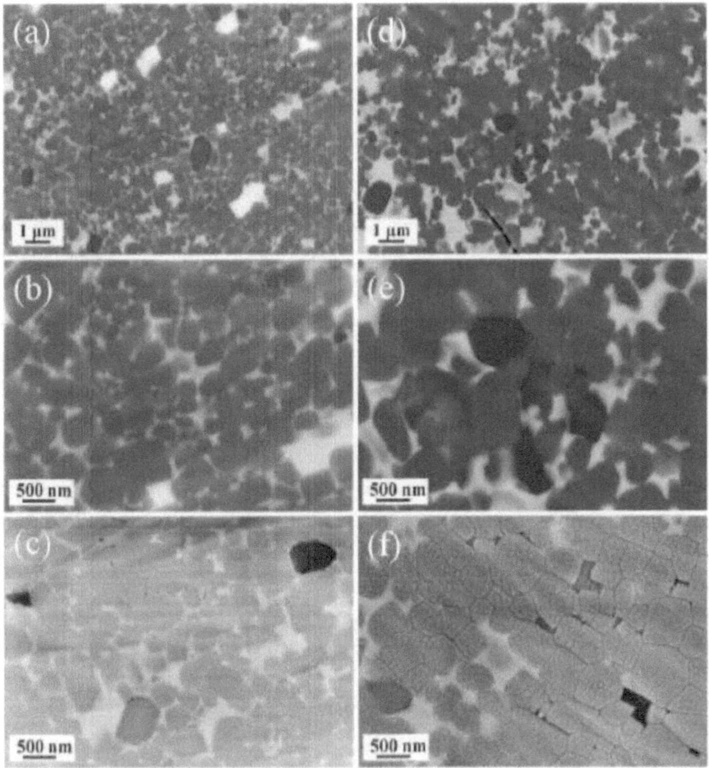

Figure 4.39: SEM images of the $(Nd_{1-x}Ce_x)_{15}Fe_{79}B_6$ HDDR powders registered on the ion-polished cross-section surface for x = 0.3 (a, b, c) and x = 0.6 (d, e, f); a, d and b, e - low and high magnification images in BSE mode; c, f - high magnification images in SE mode.

At 840 °C process temperature under controlled vacuum, desorption-recombination takes place in the presence of a rare-earth-rich liquid (from dehydrogenated intergranular $(Nd, Ce)H_x$). The rare-earth-rich liquid phase penetrates and solidifies between the $Nd_2Fe_{14}B$ grains, its redistribution ensuring the necessary grain decoupling [166, 167] to impede magnetization reversal [168] and enhance coercivity in the HDDR powders.

Observations are consistent with previous studies which indicated that the rare-earth-rich intergranular phase melts as hydrogen is desorbed during vacuum annealing at the desorption-recombination stage of the HDDR process [166, **169**]. At high Ce concentrations, in the absence of a rare-earth-rich phase, due to the decomposition of intergranular $CeFe_2H_x$ into CeH_x and α-Fe, Ce liquid phase is provided to wet the grain boundaries and form an intergranular phase between the recombined grains.

The preparation of anisotropic $Ce_{13}Fe_{80}B_7$ HDDR powders was previously reported by M. Xing et al. [141]. However, the as-treated HDDR powder showed very low coercivity and an additional processing step of intergranular infiltration with Ce-Cu eutectic alloy was necessary for improvement.

A Ce-rich over-stoichiometric alloy composition would therefore be the appropriate choice since it allows the formation of sufficient intergranular material to be decomposed and redistributed between the matrix grains during HDDR. Alternatively, grain boundary infiltration with low melting point eutectics can also help compensate for the lack of a rare-earth-rich grain boundary phase.

In the same vein, Herbst et. al [89] showed that for $Ce_2Fe_{14}B$-based rapidly quenched alloys, $Ce_{14}Fe_{79}B_7$ and $Ce_{17}Fe_{78}B_6$ compositions gave maximum magnetic properties comparing to $Nd_2Fe_{14}B$-based alloys that yield optimal properties when prepared from near stoichiometric formulations. Depending on the alloy preparation technique (be it conventional or fast solidification), the initial microstructure has to be tuned for the maximization of $Nd_2Fe_{14}B$ phase proportion. Further optimization is allowed towards rare-earth lean compositions for low Ce substitution levels and rare-earth (Ce) rich formulations in the case of high Ce content $Nd_2Fe_{14}B$ - based alloys. One important beneficial aspect to be highlighted here is that H_2 treatment facilitates the decomposition of $CeFe_2$ at low temperature whereas in normal heating conditions the compound would undergo a peritectic reaction at 925 °C decomposing into liquid and Ce_2Fe_{17} which further decomposes peritectically at 1063 °C into γ-Fe and liquid [**170**].

4.2.5 Thermal hydrogen absorption-desorption behavior

Hydrogen absorption-desorption calorimetric measurements are essential for getting a deeper insight into the processes that take place within the material during the HDDR treatment. The measurements are performed in ramping temperature, in H_2 atmosphere and in vacuum pumping conditions to show the interstitial absorption and Φ-phase matrix disproportionation reaction, and respectively the hydrogen desorption and matrix recombination reaction. In this work, the DTA measurements were performed in order to determine the influence of the Ce substitution in the strip-cast alloys on their hydrogen thermal absorption-desorption characteristics and the impact it has on their behavior during HDDR. Correlated with the SEM imaging and the XRD scans performed on the strip-cast materials at the different stages of the hydrogen treatment, the DTA results shed more light into the processes that take place during HDDR and help explain the way Ce substitution factors in the evolution of the phase structure and microstructure throughout the process.

Figure 4.40 contains DTA curves showing the sequence of thermal events occurring in the alloys with temperature ramping up to 900 °C at 10 K/min heating rate, in 1 bar H$_2$ atmosphere (Ist heating, 4.40 (a)) and in vacuum (IInd heating, run under continuous evacuation for desorption-recombination 4.40 (b)). The pressure variation with the temperature is also presented for the second heating (Figure 4.40 (c)).

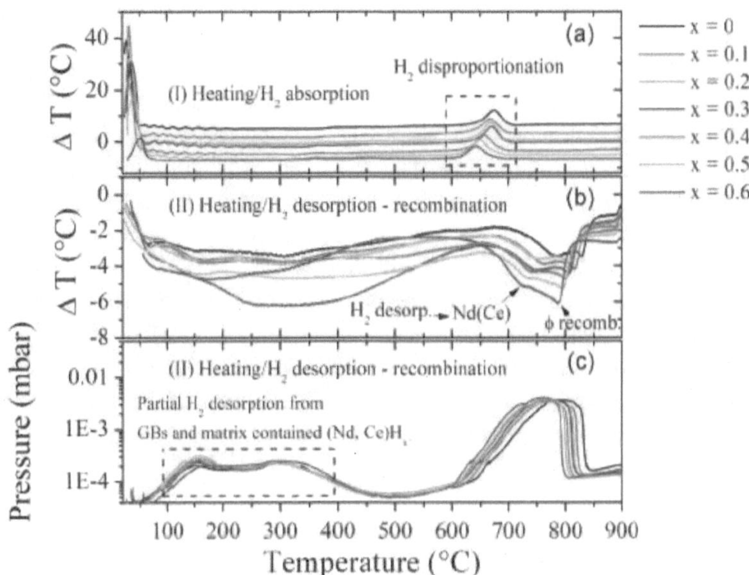

Figure 4.40: DTA scans of the (Nd$_{1-x}$Ce$_x$)$_{15}$Fe$_{79}$B$_6$ strip-cast alloys run in 1 bar H$_2$ atmosphere (Ist heating) and continuous evacuation (IInd heating) conditions.

For all samples, the Ist heating scans, run in hydrogen atmosphere, indicate two strong exothermic effects taking place marked by two peaks. The first peak situated within room temperature to ~70 °C range corresponds to the interstitial absorption of hydrogen while the peak situated above 600 °C signals the disproportionation of the hydrogenated Φ-phase matrix. In Figure 4.40 (b), for all samples, the double endothermic peaks situated at high temperature indicate consecutively the hydrogen desorption from (Nd, Ce)H$_x$ with the formation of liquid Nd(Ce) and the Φ-phase recombination as a result of Nd(Ce)$_{liq.}$ reaction with Fe$_2$B and α-Fe.

The pressure maxima registered during the vacuum heating scan signal the partial desorption of hydrogen, from the intergranular phase (the wide double peak below 500 °C, marked in Figure 4.40 (c) as grain boundary phase - GB), respectively the desorption from the (Nd, Ce)H$_x$ contained in the matrix phase disproportionated grains and the matrix recombination (the high temperature peak).

The Curie transition of the hydrogenated Φ-phase is also registered, the effect having the typical aspect of a weak exothermic step occurring on the hydrogen absorption curve (I^{st} heating, Figure 4.40 (a). Magnified detail plots of the DTA curves on the temperature interval containing the Curie transitions are introduced grouped separately for the $CeFe_2$-free alloys (at x = 0 to 0.2 Ce ratio) and for the $CeFe_2$ containing alloys (at x = 0.3 to 0.6 Ce ratio) in Figures 4.41a and 4.41b respectively. Besides the Curie transition step (indicated with arrows on all curves), additional thermal effects can be observed on the curves registered for the alloys with x = 0.3 to 0.6 Ce fractions. These are two exothermic peaks (both shifted to lower temperature for the x = 0.6 sample), the first located on the curves at the ~ 360 ... 370 °C temperatures and a second, faint peak at ~ 450 ... 460 °C, most likely indicating a stage-wise decomposition of the amorphous $CeFe_2H_x$ phase [68].

The Curie transition temperature (T_C) of the hydrogenated Φ-phase decreases monotonically with increasing Ce concentration, with a noticeably milder slope from x = 0.2 to 0.3. The DSC scans of the HDDR powders (Figures 6.11a and 6.11b in the Annex) indicated that the Curie temperatures of the recombined Φ-phase (T_C*) decrease with increasing x, following the same trend. Aside from the Ce substitution influence, due to the crystal lattice expansion (therefore larger interatomic distances) caused by the interstitial hydrogen insertion, the Curie temperature of the Φ-phase are shifted higher in hydrogenated state comparing to the non-hydrogenated phase [171]. The hydrogen-induced transformation temperatures along with the Curie temperatures of the hydrogenated Φ-phase identified on the DTA scans (T_C) in comparison to the Curie temperatures of the recombined Φ-phase derived from DSC scans (T_C*) are listed in Table 4.8.

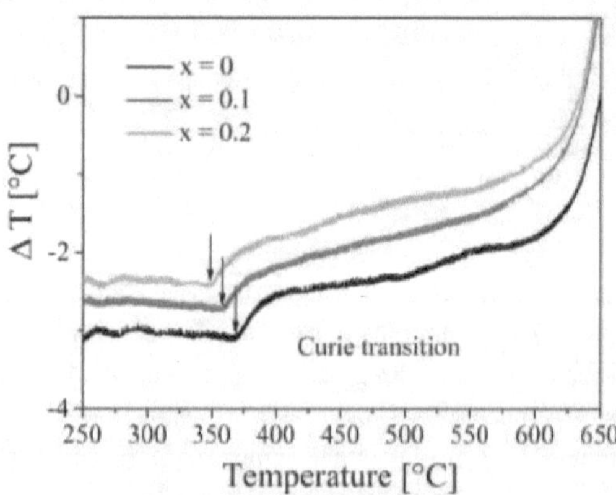

Figure 4.41a: Magnified detail on the DTA scans run in 1 bar H_2 atmosphere (I^{st} heating curves, in Figure 4.40 (a)) showing the decrease of hydrogenated Φ-phase Curie temperatures with increasing Ce fractions in the $(Nd_{1-x}Ce_x)_{15}Fe_{79}B_6$ strip-cast alloys for x = 0, 0.1 and 0.2.

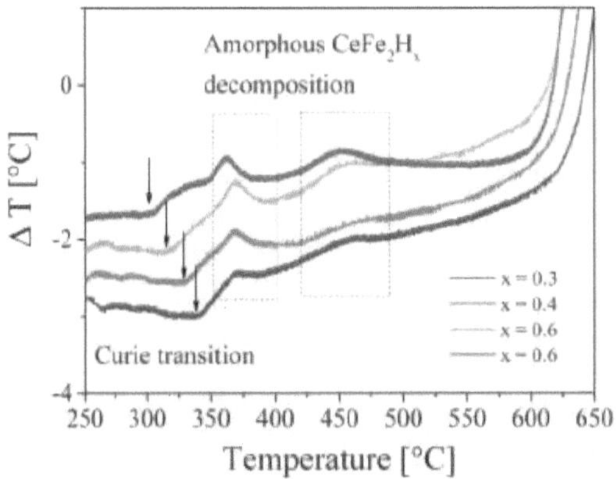

Figure 4.41b: Magnified detail (from I[st] heating curves, in Figure 4.40 (a)) showing the decrease of the hydrogenated Φ-phase Curie transition temperatures with increasing Ce fractions on the DTA scans of the $(Nd_{1-x}Ce_x)_{15}Fe_{79}B_6$ strip-cast alloys that contain $CeFe_2$ (at x = 0.3 to 0.6).

Table 4.8: Hydrogen absorption peak temperatures ($T_{H\ abs.\ peak}$), Curie temperatures of the hydrogenated Φ-phase (T_C), disproportionation onset ($T_{disp.\ onset}$) and peak temperatures ($T_{disp.\ peak}$), hydrogen desorption ($T_{desorp.\ peak}$) and Φ-phase recombination peak temperatures ($T_{\Phi\ recomb.\ peak}$) determined from DTA scans on the $(Nd_{1-x}Ce_x)_{15}Fe_{79}B_6$ strip-cast alloys. The Curie temperature of the recombined Φ-phase (T_C^* notation) was derived from DSC scans run on the fully-processed HDDR powders.

X	$T_{H\ abs.\ peak}$	T_C	T_C^*	$T_{disp.\ onset}$	$T_{disp.\ peak}$	$T_{desorp.\ peak}$	$T_{\Phi\ recomb.}$
	(°C)						
x = 0	26.77	369.52	310.15	640.99	675.13	786.31	828.99
x = 0.1	31.00	358.85	299.80	631.08	668.20	760.39	815.12
x = 0.2	34.19	348.47	288.45	640.05	668.56	750.48	809.89
x = 0.3	35.07	341.67	281.53	638.32	670.72	747.26	806.27
x = 0.4	30.67	329.80	268.70	626.16	658.57	741.60	800.53
x = 0.5	39.50	319.67	254.11	616.48	648.77	735.60	794.67
x = 0.6	51.96	305.06	234.63	612.11	642.25	726.31	786.28

The hydrogen absorption peak temperatures ($T_{H\ abs.\ peak}$) show a non-monotonic increase with increasing Ce concentrations in the strip-cast alloys composition. Hydrogen diffuses in the material interstitially, through the rare-earth-rich grain boundary phase first then entering the Φ-phase matrix. As shown by XRD measurements and EPMA mapping (shown in Section 4.2.1), the phase structure of the intergranular material changes with Ce concentration in the strip-cast alloys.

The rise of $T_{H\ abs.\ peak}$ values is therefore associated with the presence of Ce mixed in the Nd-rich phase and the appearance of $CeFe_2$ phase segregations at the grain boundaries. The absorption peak temperature is highest ($T_{H\ abs.\ peak}$ = 51.96 °C) at x = 0.6 Ce fraction where the intergranular material is composed mostly of $CeFe_2$ which transforms in amorphous $CeFe_2H_x$ at this stage (shown in the analysis of decrepitated alloys - section 4.2.2).

The DTA data also shows that, as Ce increasingly substitutes Nd in the Φ-phase structure, the disproportionation reaction is shifted to lower temperatures. At x = 0.1, Ce is incorporated into the $Nd_2Fe_{14}B$ matrix phase structure decreasing the disproportionation temperature comparing to the Ce-free sample. At x = 0.2 and 0.3, Ce is more concentrated in the intergranular material (mixed within the Nd-rich phase and also forms $CeFe_2$ (at x = 0.3 Ce ratio)). Now the disproportionation onset temperature ($T_{disp\ onset}$) increases becoming comparable to the Ce-free sample while the peak temperature ($T_{disp.\ peak}$) remains stable at x = 0.2 (relative to x = 0.1) and increases slightly at x = 0.3. At x = 0.4 and higher, more Ce atoms substitute Nd within the Φ-phase structure, steadily decreasing the disproportionation temperature. The hydrogen absorption temperature shows a non-monotonic increase with increasing x (it is conditioned by the initial phase structure of the intergranular material), while the desorption temperature decreases monotonically. The desorption temperature monotonic decrease is due to the removal of initial diffusion barriers at the intergranular regions as direct diffusion paths are created by $CeFe_2H_x$ decomposition into CeH_x and α-Fe.

4.2.6 Magnetic properties of HDDR powders

Figures 4.42a and 4.42b contain the room temperature hysteresis loops of the $(Nd_{1-x}Ce_x)_{15}Fe_{79}B_6$ HDDR powders (x = 0, 0.1, 0.2 .. 0.6), measured parallel (easy axis) and perpendicular to the magnetization direction (hard axis) and a plot of the coercivity (μ_0H_C) and the easy axis remanence ($B_r^{|}$) against Ce concentration. The magnetic properties (values of coercivity μ_0H_C, remanence B_r parallel and perpendicular to the easy axis and degree of texture (DOT)) of the prepared HDDR powders are summarized in Table 4.9. The step visible at roughly 0.28 T field on the demagnetization curves in the second quadrant for the HDDR powders with x = 0.1, 0.4 and 0.5 Ce concentrations signals the magnetization reversal of a secondary phase with low coercivity, most probably residual α-Fe or, alternatively, a separated Φ-phase with higher Ce content. The possible separated Φ-phase is suggested by a lower temperature Curie transition on the DSC traces of the HDDR powders (the related data is annexed in Section 6, in Figure 6.11b. Strong texture in the Ce containing alloys was obtained at x = 0.2 and 0.3, compositions which also showed stable values of coercivity and remanence. The coercivity (μ_0H_C) shows a monotonic decrease with increasing Ce substitution ratio. The remanence ($B_r^{|}$) shows a non-monotonic decreasing trend as Ce content is increased, with a steep decline at x = 0.1 Ce concentration.

Table 4.9: Magnetic properties of $(Nd_{1-x}Ce_x)_{15}Fe_{79}B_6$ HDDR powders - coercivity μ_0H_C, remanence B_r parallel and perpendicular to the magnetic easy axis and degree of texture (DOT). DOT was defined as the value of $(B_r^{\parallel} - B_r^{\perp})/B_r^{\parallel} *100$.

Ce at. ratio	x = 0	x = 0.1	x = 0.2	x = 0.3	x = 0.4	x = 0.5	x = 0.6
μ_0H_c (T)	1.3	1.16	1.07	1.05	0.99	0.83	0.67
$B_r^{\parallel}$ (T)	0.97	0.75	0.89	0.88	0.74	0.70	0.66
$B_r^{\perp}$ (T)	0.55	0.61	0.59	0.54	0.64	0.54	0.51
DOT (%)	43	19	34	39	14	23	23

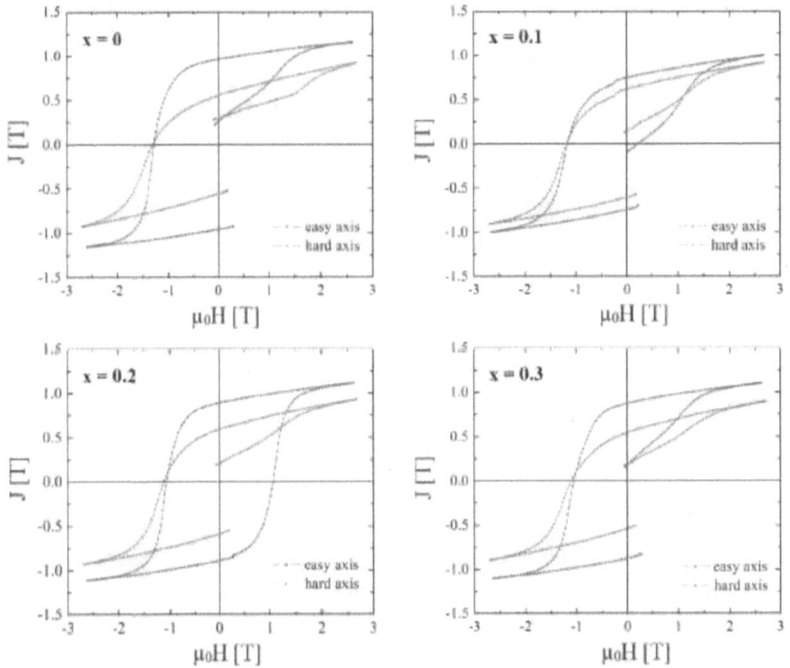

Figure 4.42a: Room temperature hysteresis loops registered on the easy and hard magnetic axis for the $(Nd_{1-x}Ce_x)_{15}Fe_{79}B_6$ (x = 0, 0.1, ... 0.3) HDDR powders.

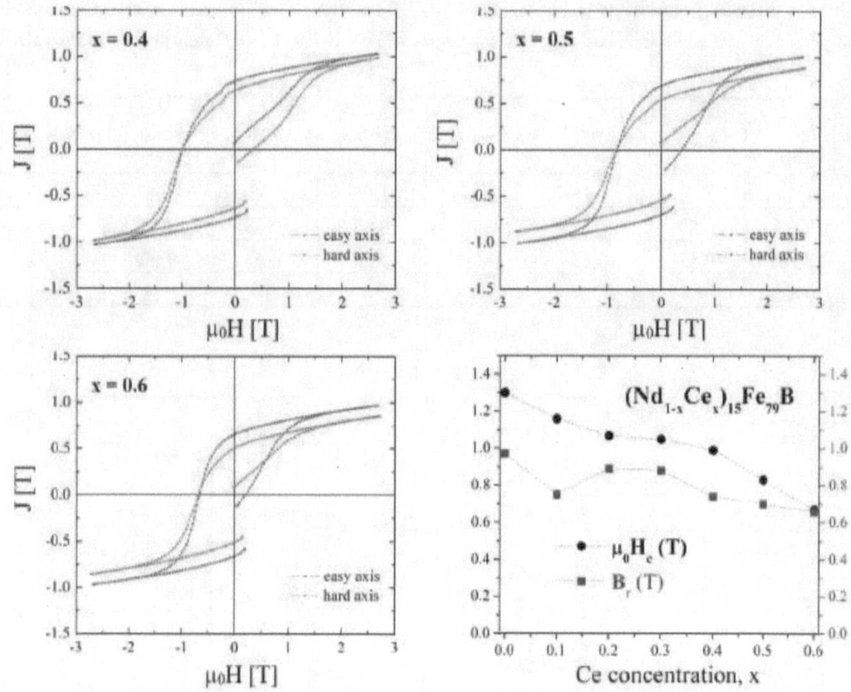

Figure 4.42b: Room temperature hysteresis loops measured on the easy and hard magnetic axis for the $(Nd_{1-x}Ce_x)_{15}Fe_{79}B_6$ (x = 0.4, 0.5 and 0.6) HDDR powders together with a plot of the coercivity μ_0H_C and the easy axis remanence B_r versus Ce fraction for all HDDR samples.

5 Summary, conclusions and perspectives

Aiming to address the resource-criticality of rare-earth raw materials necessary for the production of rare-earth permanent magnets, this book analyzes the utilization of Ce and La as affordable rare-earth elements for the substitution of Nd in $Nd_2Fe_{14}B$-based alloys. Two processing routes were considered as two related case studies to illustrate the effects of substitution. In brief, this work contains experimental research data on the structural and magnetic properties of Ce and La substituted $Nd_2Fe_{14}B$-based alloys produced by melt-spinning and hot-working and, in a complementary frame of analysis, on the effects of Ce substitution in strip-cast and hydrogen treated alloys. The influence of the alloy composition and processing conditions on the phase solidification, crystal structure, microstructure and resulting magnetic properties was analyzed for the two adopted processing routes using various material investigation techniques such as X-ray powder diffraction, electron microscopy imaging, calorimetric scans, magnetic hysteresis tracing and thermomagnetic measurements.

As a first case study in this work, the substitution of Ce and La for Nd in nanocrystalline melt-spun ribbon alloys of nominal compositions $(Nd_{1-x}R_x)_{13.6}Fe_{bal}Co_{6.6}Ga_{0.6}B_{5.6}$ ($x = 0, 0.1, ... 1$ for R = Ce and $x = 0, 0.1, ... 0.5$ for R = La) was analyzed. The XRD analysis on crystal properties and phase composition showed that Ce substitution gradually decreased the $Nd_2Fe_{14}B$ phase lattice constants and unit cell tetragonality while inducing $CeFe_2$ Laves-type phase segregations at high Ce fractions (starting at $x = 0.7$). SEM analysis showed that grain growth was induced even at low Ce concentration, with a significantly coarse microstructure observed in the $Ce_{13.6}Fe_{bal}Co_{6.6}Ga_{0.6}B_{5.6}$ melt-spun ribbons. The magnetic properties of the melt-spun powders decreased with increasing Ce fractions.

For $(Nd_{1-x}La_x)_{13.6}Fe_{bal}Co_{6.6}Ga_{0.6}B_{5.6}$ melt-spun alloys, analysis of the XRD patterns showed that La induced lattice expansion along the c-axis. The alloys are single-phase up to $x = 0.4$ and contain primary α-Fe and Nd_2Fe_{17} phase segregations at $x = 0.5$. Lanthanum substitution refined the microstructure suppressing the grain growth. The coercivity decreased with increasing La concentration but the remanent magnetization increased from 0.73 T in the reference composition to 0.88 T at $x = 0.3$ La concentration. Also for moderate concentrations, La substitution for Nd did not noticeably affect the saturation magnetization as a result of its tendency to migrate outside the main phase, to the grain boundaries. For this reason as well, a high Curie temperature was maintained which could also be due to the anisotropic lattice expansion produced by the larger La atom replacing Nd in the $Nd_2Fe_{14}B$ crystal lattice. The observed remanence enhancement with La substitution in the melt-spun alloys is due to slightly increased fractions of the $Nd_2Fe_{14}B$ phase, this because its formation is promoted as La does not form compounds with Fe.

For the alloy compositions containing $x = 0.2$ and 0.3 Ce and $x = 0.2$ La fractions, the melt-spun powders were used for magnet fabrication by hot-pressing and hot-deformation. The hot-pressed $(Nd_{0.8}La_{0.2})_{13.6}Fe_{bal}Co_{6.6}Ga_{0.6}B_{5.6}$ alloy measured lower coercivity but increased remanence relative to the Ce-substituted alloys.

However, this composition responded poorly to hot-deformation, severe cracking being produced in the process, presumably caused by the fact that La raises the melting temperature of the intergranular eutectic phase. Also the possible presence of lanthanum oxides at the grain boundaries could contribute to the sample fracture. The alloy processing into anisotropic hot-deformed magnets showed that Ce substitution brings about a technological benefit, enabling hot-workability and grain deformation by decreasing the melting temperature of the rare-earth-rich phase which results in higher texture and implicitly in a higher magnetization and magnetic energy density product. Due to enhanced deformability, best magnetic properties were obtained after deformation for the $(Nd_{0.7}Ce_{0.3})_{13.6}Fe_{bal}Co_{6.6}Ga_{0.6}B_{5.6}$ alloy which measured the coercivity $\mu_0H_c = 1.09$ T, the remanence $B_r = 0.97$ T and the magnetic energy density product $|BH|_{max} = 170$ kJ/m^3.

As a second case study, the influence of Ce substitution on the phase structure, microstructure and the related magnetic property development in hydrogen decrepitated and HDDR processed $(Nd_{1-x}Ce_x)_{15}Fe_{79}B_6$ (x = 0, 0.1, ..., 0.6) strip-cast alloys was analyzed. Intergranular $CeFe_2$ phase was found to segregate in the as-cast alloys starting at x = 0.3 Ce concentration. It increases in proportion with x at the expense of the hard magnetic Φ-phase, replacing the rare-earth-rich grain boundary phase. Upon hydrogen absorption during decrepitation, $CeFe_2$ transforms in amorphous $CeFe_2H_x$ which further decomposes into CeH_x and α-Fe upon heating during HDDR. These findings come in agreement with data reported in the literature describing the behavior of the $CeFe_2$ compound under hydrogen exposure conditions. The rare-earth element components from CeH_x and the rare-earth-rich hydrogenated phase melt upon hydrogen desorption. The melting enables the redistribution of the intergranular material among the $Nd_2Fe_{14}B$ grains upon the desorption and recombination stage of the HDDR treatment, thus facilitating grain decoupling and coercivity development in the HDDR powders. Reasonable levels of coercivity are attained even for the alloys with high Ce content in which the intergranular phase is composed mainly of $CeFe_2$ showing that, similarly to the Nd-rich phase, $CeFe_2$ can effectively decouple the grains. The redistribution of the $CeFe_2$ phase during the process is of key importance in coercivity development along with the beneficial effect of grain refinement brought about by HDDR treatment.

The magnetic properties of the HDDR powders showed a decreasing trend with increasing Ce concentration. However at x = 0.2 and 0.3 Ce ratios, coercivity (μ_0H_C) and remanence (B_r) of the HDDR powders remained rather stable despite the segregation of $CeFe_2$ phase occurring at x = 0.3. Significant texture and optimal magnetic properties were obtained at x = 0.3 ($\mu_0H_C = 1.05$ T and $B_r = 0.88$ T) attributed to a favorable phase distribution with Ce mainly concentrated in the intergranular material. Differential thermal analysis showed that the hydrogen thermal absorption-desorption behavior of the alloys (described in this work through hydrogen absorption, disproportionation and recombination temperatures variation with Ce concentration) is sensitive to the phase composition and the distribution of Ce between the matrix and the intergranular phases.

The maximization of magnetic properties of the Ce and La-substituted magnets and $Ce_2Fe_{14}B$ magnets calls for microstructure engineering by correlating the alloy composition and processing or magnet production routes. Microstructures free of segregated phases, with fine, uniformly sized grains (size scale and morphology depend on solidification method) separated by a thin and uniformly distributed non-magnetic intergranular phase are generally required.

For isotropic bonded magnets fabrication, further optimization of the melt-spinning process parameters for the specific alloy composition for a more uniform and fine-grained microstructure of the melt-spun flakes should lead to improved magnetic properties.

The use of element additions for the grain refinement of Ce-substituted melt-spun alloys to further increase coercivity can be tested (an example that proved to have this effect is La). Also the use of inoculants for grain refinement can also be tested (refractory powders such as ZrC used by Pathak et al. [90], that could act as nucleation sites during solidification or restrict growth during thermal treatment for the crystallization of amorphous ribbons). Testing could be done by DSC investigation for the undercooling degree variation by following the melting and solidification characteristics with increasing cooling rate, additive type and concentration.

Production routes by thermal (sintering) and thermo-mechanical powder densification (hot-pressing and hot-deformation) come with further requirements regarding the alloy's processability. In these processing scenarios the properties of the intergranular phase are decisive.

Again in this work, the HDDR reaction was proved to be also a powerful probing method which provides very useful information, allowing the gain of new insights also for the powder sintering of $Ce_2Fe_{14}B$-based alloys. One highlight of this work is that hydrogenation allows the decomposition of $CeFe_2$ at low temperature (this could facilitate the low temperature sintering) whereas in normal heating conditions the compound decomposes through peritectic reaction at 925 °C into liquid and Ce_2Fe_{17} which further undergoes a peritectic decomposition into liquid and γ-Fe at 1063 °C.

Further experimental work could link the hydrogen absorption to the milling properties of the Ce-substituted alloys containing larger amounts of intergranular $CeFe_2$ as these alloys are noticeably less brittle comparing to the $Nd_2Fe_{14}B$ ones also in hydrogen decrepitated condition. Also further work could be done for the assessment (if not control) of the oxidation behavior of the alloys keeping in mind that the condition of the rare-earth component of the intergranular material (its chemical nature and whether it is in a compound or in an eutectic mixture) is a determining factor.

Considering the liquid phase formation at high temperature through the aforementioned sequence of peritectic decompositions, the presence of $CeFe_2$ phase in $Ce_2Fe_{14}B$-based melt-spun alloys is an impediment in the liquid phase assisted hot-pressing and hot-deformation processing which for $Nd_2Fe_{14}B$-based alloys is typically conducted above the melting point of the Nd-rich phase, at temperatures around 750 °C. The strategy of compensating through grain boundary infiltration with low melting eutectics mixtures is therefore suitable and a necessity in such a processing scenario. For instance infiltration with low melting eutectics having a diamagnetic and a rare-earth component would not only create an intergranular phase to decouple the grains (considering that $CeFe_2$, unlike the Nd-rich eutectic phase, is discontinuous at the grain boundaries) but through melting it would also provide the liquid rare-earth component to diffuse and substitute Ce in the Φ-phase at the grain surface and create core-shell structured grains with higher anisotropy at the shell regions.

Additive elements can be tested such as Si which is known to partially substitute for Fe [33], slightly raise T_C (presumably due to lattice expansion as Si is larger than Fe) and improve remanence in $Nd_2Fe_{14}B$ alloys [**172**].

Silicon was also shown to improve the coercivity of Ce-substituted magnets while keeping T_C relatively stable [106] which suggests that the element is partitioned between the hard magnetic phase (hence the slight effect on T_C) and the intergranular material (with effects on the resulting microstructure (improved hot-workability [159]) hence the effect on coercivity). The search for additives that would impede grain coarsening also during thermal treatment and hot-compaction (for example to create refractory precipitates such as ZrC and TiC [90, **173**]) would lead to higher coercivities.

Another potential research direction is the co-substitution of Ce and La that was very recently shown to help promote the formation of the Φ-phase and suppress the formation of the $CeFe_2$ phase which would further call for more insight into grain boundary processes and response to various processing conditions (for example oxidation behavior, hydrogen absorption-desorption properties, response to decrepitation treatment and related degree of embrittlement, powder sinterability, deformability of melt-spun alloys).

For coercivity enhancement through anisotropy increase at the grain surface, the novel approach on grain boundary diffusion (GBDP) of rare-earths (for instance Dy or Tb) from sputtered thin layers as suggested by recent literature would be not only a clean and environment-friendly alternative but also a resource-efficient one because it maximizes the material consumption efficiency.

The findings in this work open potentially interesting paths for further research, particularly in solidification and grain boundary phase related processes for it is the phenomena occurring during these processes that provide opportunities for deeper understanding and creative changes in the design of alloy microstructures and for the optimization of the magnetic properties of permanent magnets containing abundant and inexpensive rare-earths.

6 Appendix

$(Nd_{1-x}La_x)_{12.5}Fe8_{1.2}B_{6.3}$ strip-cast alloys - phase structure and hydrogen absorption-desorption characteristics

Figures 6.1 contains the XRD patterns of the alloys of composition $(Nd_{1-x}La_x)_{12.5}Fe_{81.2}B_{6.3}$ (x = 0, 0.1 ... 0.3) produced by strip-casting. The unlabeled peaks correspond to the Φ-phase. The primary α-Fe segregated phase (which becomes more significant in proportion at x = 0.2 and 0.3 La fractions) and the minor η-boride phase are marked on the patterns.

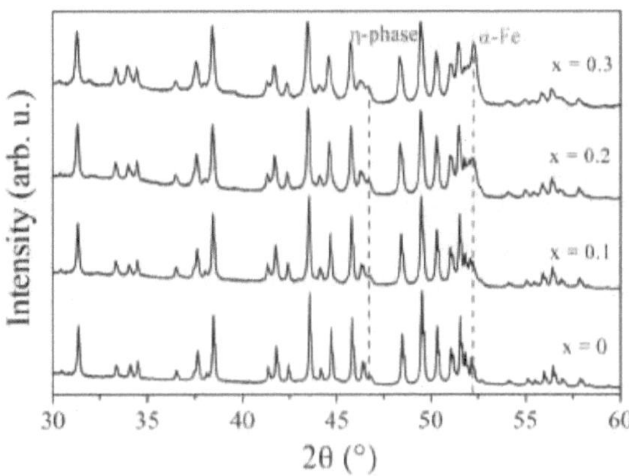

Figure 6.1: XRD scans of the $(Nd_{1-x}La_x)_{12.5}Fe_{81.2}B_{6.3}$ (x = 0, 0.1 ... 0.3) strip-cast alloys.

DTA hydrogen absorption-desorption curves measured on $(Nd_{1-x}La_x)_{12.5}Fe_{81.2}B_{6.3}$ strip-cast alloys (for x = 0, 0.1 and 0.2) are shown in Figures 6.2, 6.3 and 6.4. On the absorption-disproportionation curves in Figure 6.2 it can be seen that increasing La substitution fractions induce the decrease of the interstitial hydrogen absorption temperatures. One may observe also that the Φ-phase disproportionation temperatures decrease with increasing La fractions in the alloy. This indicates that the composition of the matrix phase is changed upon La substitution for Nd, confirming the retention of La atoms its structure. However due to lattice expansion, the Curie transition temperature of the hydrogenated Φ-phase remains stable at x = 0.1 La and is only slightly decreased at x = 0.1 La. The data also shows that the Φ-phase recombination temperature does not vary significantly from x = 0 to 0.1 La fraction (possibly indicating that La diffuses at the grain boundaries at the desorption-recombination stage of the process) and is decreased for the sample with x = 0.2 La concentration.

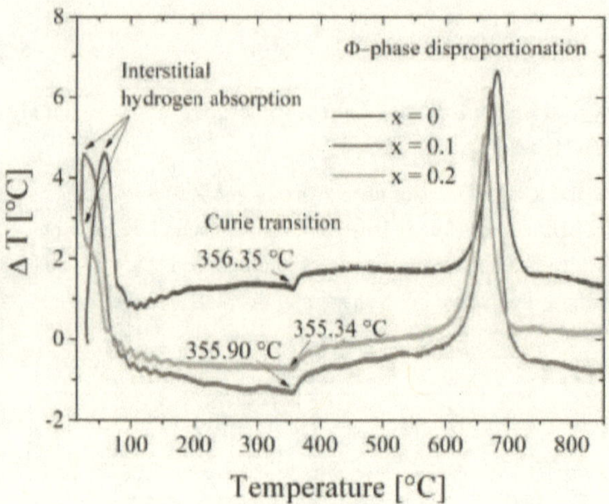

Figure 6.2: DTA scans of the $(Nd_{1-x}La_x)_{12.5}Fe_{81.2}B_{6.3}$ (x = 0, 0.1 and 0.2) strip-cast alloys run in 1 bar H_2 atmosphere and at 10 K/min heating rate. The Curie transition temperatures marked on the curves corresponds to the hydrogenated Φ-phase.

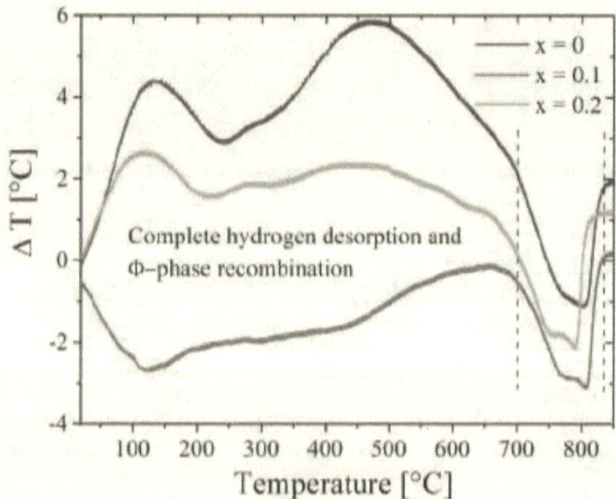

Figure 6.3: DTA scans of the $(Nd_{1-x}La_x)_{12.5}Fe_{81.2}B_{6.3}$ (x = 0, 0.1 and 0.2) disproportionated alloys run at 10 K/min heating rate under continuous vacuum pumping. The temperature interval of interest containing the desorption-recombination events is marked in the figure.

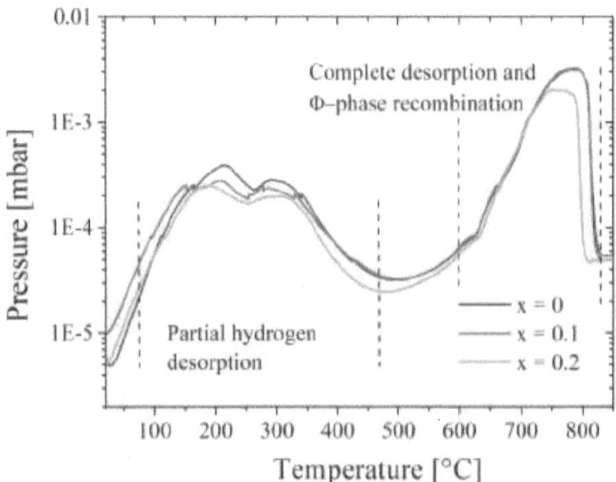

Figure 6.4: The pressure versus temperature variation registered during the DTA scans of the $(Nd_{1-x}La_x)_{12.5}Fe_{81.2}B_{6.3}$ (x = 0, 0.1 and 0.2) disproportionated alloys run at 10 K/min heating rate under continuous vacuum pumping.

The temperature intervals marked in the figure contain the following thermal events, in consecutive order: the partial desorption of hydrogen from grain boundaries and matrix phase rare-earth hydrides signaled by the wide double peaks and the complete desorption and matrix recombination signaled by the pressure maxima situated at high temperature.

M(T) curves of $(Nd_{1-x}R_x)_{13.6}Fe_{bal}Co_{6.6}Ga_{0.6}B_{5.6}$ melt-spun alloys

Figures 6.5a, 6.5b and 6.6 present the M(T) curves measured on the melt-spun alloys of composition $(Nd_{1-x}R_x)_{13.6}Fe_{bal}Co_{6.6}Ga_{0.6}B_{5.6}$ at. % (x = 0, 0.1, 0.2, ... 1 for R = Ce and x = 0, 0.1, 0.2, ... 0.5 for R = La). The derived T_C values (taken as the minima on the first derivative curves are also introduced and discussed Chapter 4, Section 4.1.3).

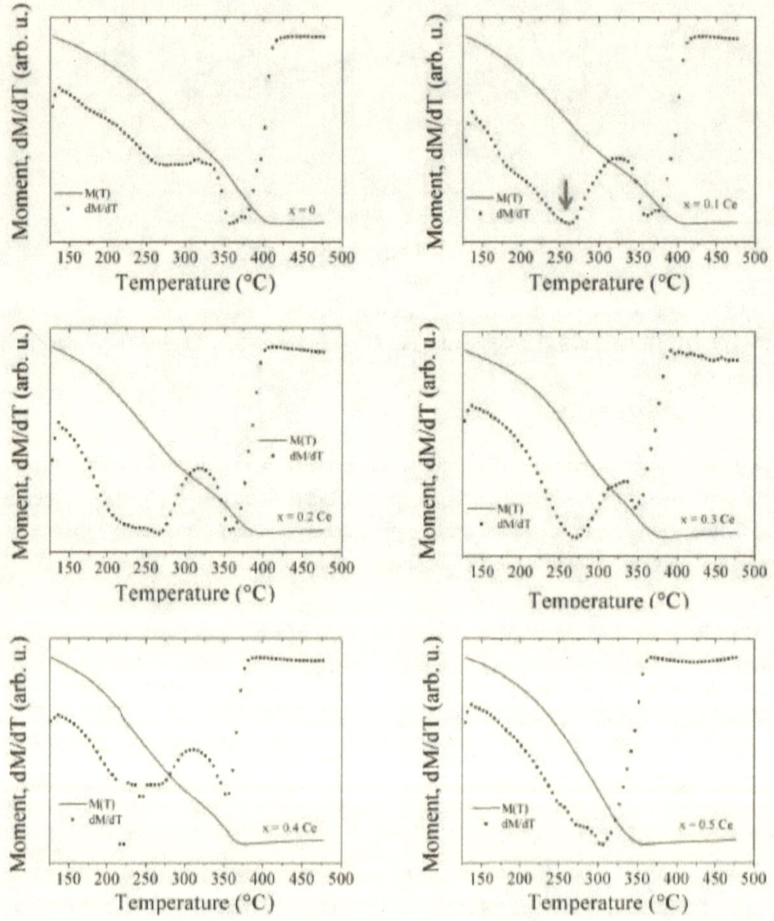

Figure 6.5a: M(T) and dM/dT curves of the $(Nd_{1-x}Ce_x)_{13.6}Fe_{73.6}Co_{6.6}Ga_{0.6}B_{5.6}$ at. % as-quenched melt-spun alloys (x = 0, 0.1, ... 0.5).

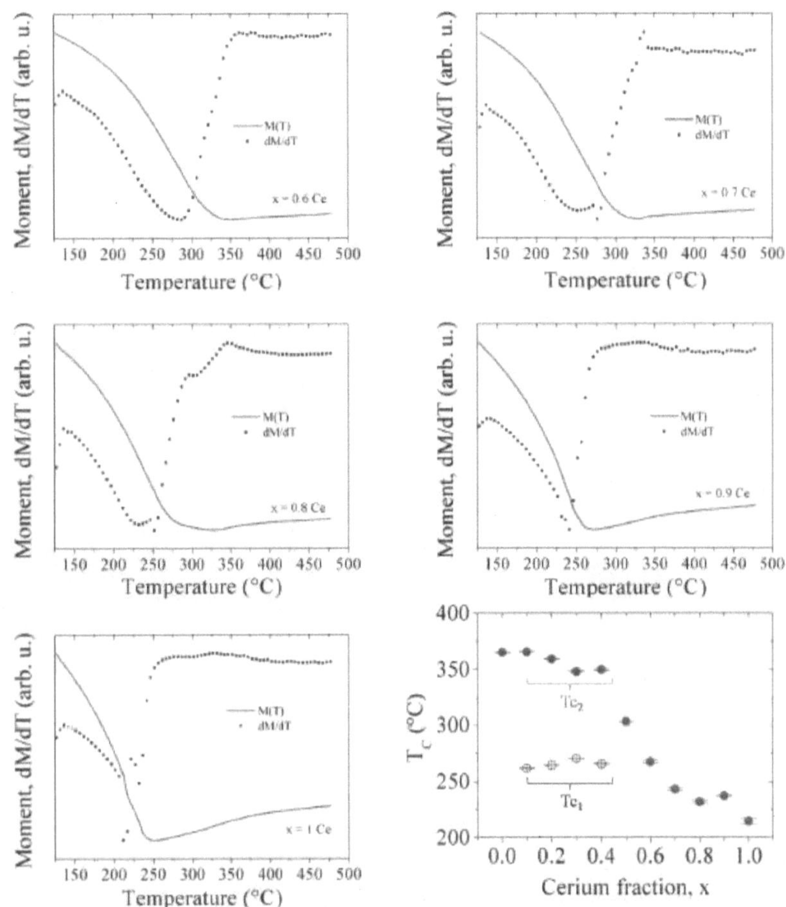

Figure 6.5b: M(T) and dM/dT curves of the $(Nd_{1-x}Ce_x)_{13.6}Fe_{73.6}Co_{6.6}Ga_{0.6}B_{5.6}$ at. % as-quenched melt-spun alloys (x = 0.6, 0.7, … 1) with the plot of T_c against the Ce fraction.

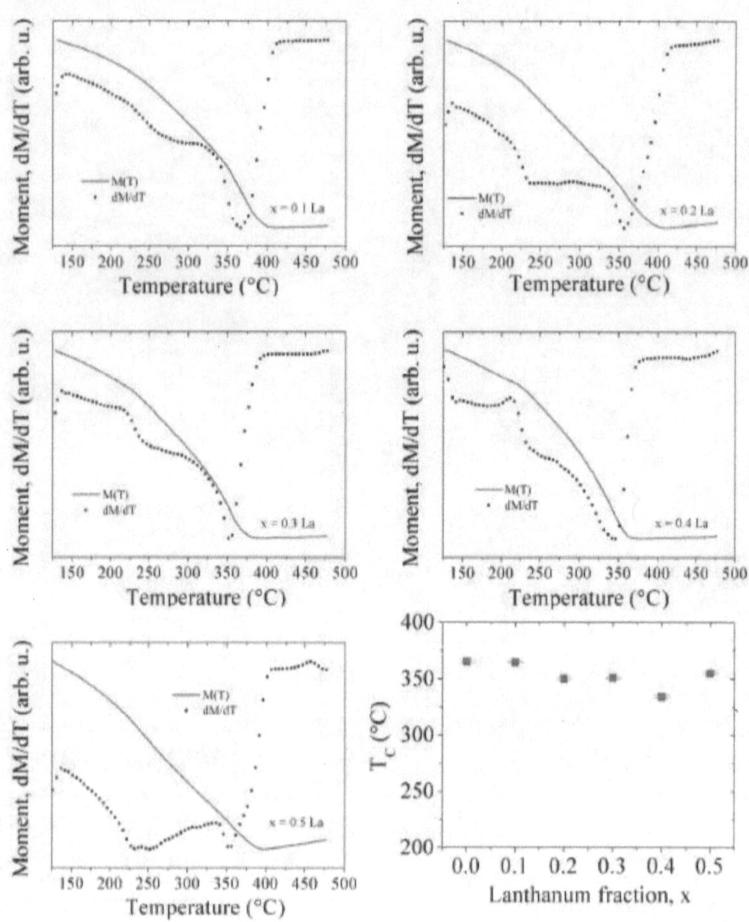

Figure 6.6: M(T) and dM/dT curves of the $(Nd_{1-x}Ce_x)_{13.6}Fe_{73.6}Co_{6.6}Ga_{0.6}B_{5.6}$ at. % as-quenched melt-spun alloys (x = 0.1, 0.2, ... 0.5) with the plot of T_C against the La fraction.

BSE images of $(Nd_{1-x}Ce_x)_{15}Fe_{79}B_6$ alloys in as-cast and disproportionated state

The quantified elemental concentrations measured and averaged over the points marked in the images in Figure 6.7 to 6.10 are listed in Table 4.6 in Section 4.2.4.

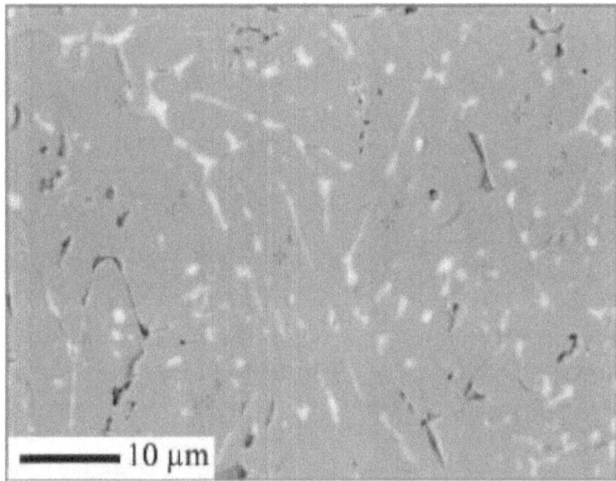

Figure 6.7: BSE micrograph of the cross-section surface of a strip-cast flake with $Nd_{15}Fe_{79}B_6$ nominal composition.

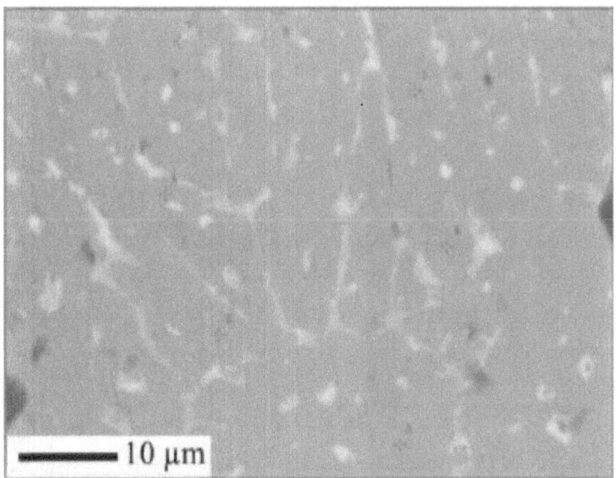

Figure 6.8: BSE micrograph on the cross-section surface of a strip-cast flake with nominal composition $(Nd_{1-x}Ce_x)_{15}Fe_{79}B_6$ for x = 0.3 Ce concentration.

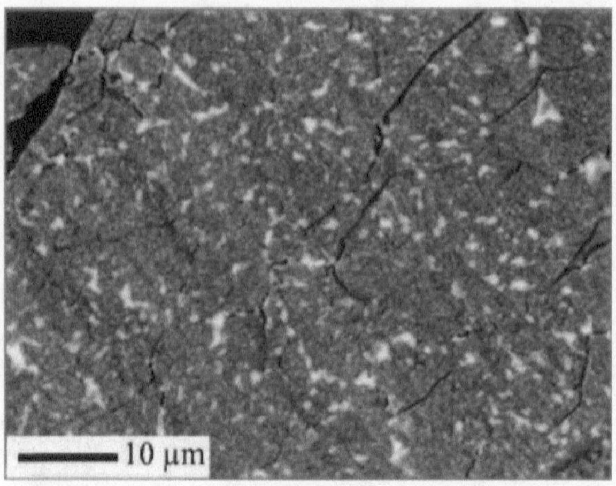

Figure 6.9: BSE micrograph of the cross-section surface of a disproportionated powder particle obtained from the strip-cast alloy of $Nd_{15}Fe_{79}B_6$ at. % nominal composition.

Note: Point 1 was excepted form the calculated average composition in Table 4.4 because it is very close to a Nd-rich region, measuring a much higher Nd concentration.

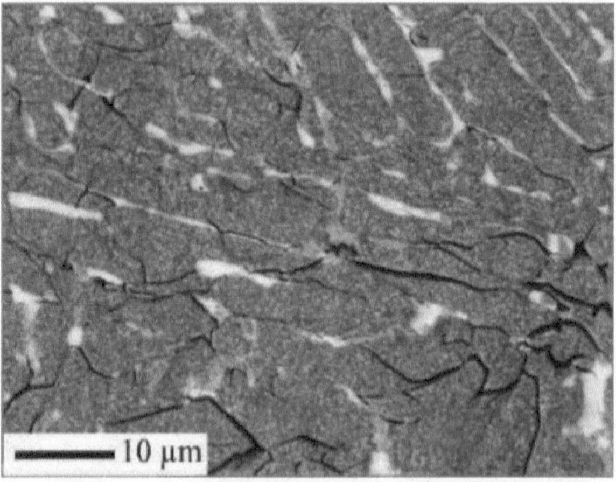

Figure 6.10: BSE micrograph of the cross-section surface of a disproportionated powder particle obtained through processing of the strip-cast alloy with $(Nd_{1-x}Ce_x)_{15}Fe_{79}B_6$, x = 0.3 composition.

DSC curves of the $(Nd_{1-x}Ce_x)_{15}Fe_{79}B_6$ HDDR powders

The DSC scans run to identify the Φ-phase Curie transition temperatures for the $(Nd_{1-x}Ce_x)_{15}Fe_{79}B_6$ HDDR powders are introduced in Figure 6.11a. The derived T_C values (shown on the first derivative curves in Figure 6.11b) are also introduced in Table 4.6 (Chapter 4, Section 4.2.5) for comparison with the T_C values corresponding to the hydrogenated Φ-phase extracted from DTA data.

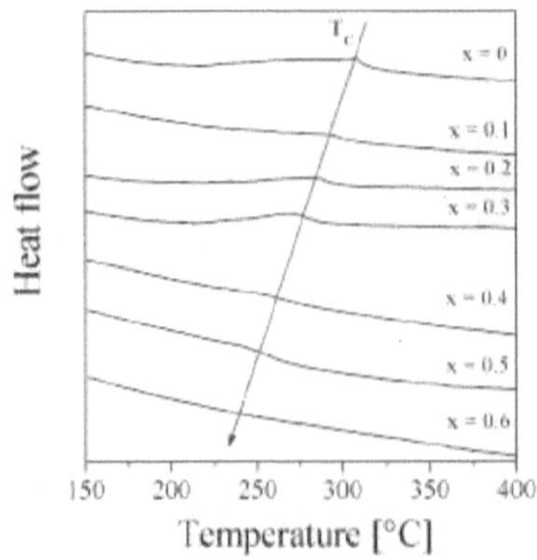

Figure 6.11a: DSC traces of the $(Nd_{1-x}Ce_x)_{15}Fe_{79}B_6$ HDDR powders run at 10°C /min heating rate (exothermic direction downward).

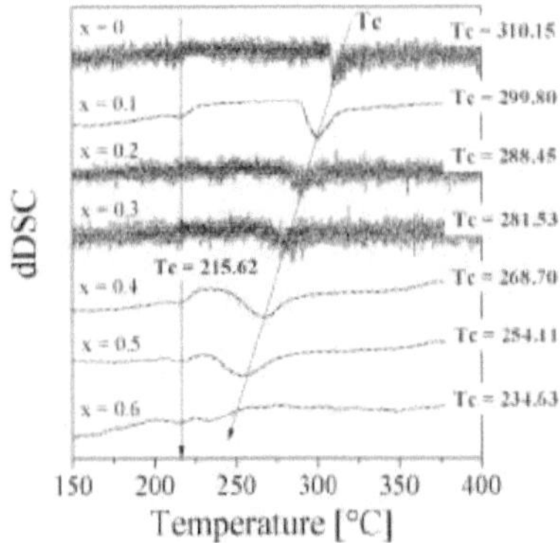

Figure 6.11b: dDSC curves of the $(Nd_{1-x}Ce_x)_{15}Fe_{79}B_6$ HDDR powders. In blue color a possible Curie transition of a separated Ce-rich Φ-phase.

9 783384 245915